66 Structure and Bonding

Electronegativity

Editors: K. D. Sen and C. K. Jørgensen

With Contributions by
J. A. Alonso L. C. Balbás L. J. Bartolotti D. Bergmann
M. C. Böhm M. Galván J. L. Gázquez J. Hinze
W. J. Mortier J. Mullay P. C. Schmidt K. D. Sen A. M. Vela

With 39 Figures and 36 Tables

Springer-Verlag
Berlin Heidelberg GmbH

Guest Editor

Professor *Kali Das Sen,* School of Chemistry, University of Hyderabad, Hyderabad 500134, India

Library of Congress Cataloging Publication Data Electronegativity. (Structure and bonding ; 66) Bibliography: p 1 Electronegativity I Sen, K. D. (Kali Das), 1948– II Jørgensen, Christian Klixbull III Alsono, J. A. (Julio A.), 1948– IV Series QD461.S92 Vol. 66 541 2'2 s 87-16401

DOI 10.1007/978-3-540-47786-0

Originally published by Springer-Verlag Berlin Heidelberg New York in 1987
MyCopy version of the original edition 1987

Typesetting: Mitterweger Werksatz GmbH, 6831 Plankstadt, Germany

2151/3140-543210
www.springer.com/mycopy

Editor's Note

Electronegativity, perhaps the most popular intuitive concept in chemistry, can now be treated as a quantum chemical parameter. This volume is an attempt to record the developmental phase of research activity in this area.

K. D. Sen

Table of Contents

Estimation of Atomic and Group Electronegativities
J. Mullay 1

Absolute Electronegativities as Determined from Kohn-Sham Theory
L. J. Bartolotti 27

Simple Density Functional Theory of the Electronegativity and Other Related Properties of Atoms and Ions
J. A. Alonso, L. C. Balbás 41

Fukui Function, Electronegativity and Hardness in the Kohn-Sham Theory
J. L. Gázquez, A. M. Vela, M. Galván 79

Electronegativity of Atoms and Molecular Fragments
K. D. Sen, M. C. Böhm, P. C. Schmidt 99

Electronegativity Equalization and its Applications
W. J. Mortier 125

Electronegativity and Charge Distribution
D. Bergmann, J. Hinze 145

Author Index Volumes 1–66 191

Estimation of Atomic and Group Electronegativities

John Mullay

Atlas Powder Company, Chemistry Section, Subsidiary of Tyler Corporation, Atlas Research and Development Laboratory, P.O. Box 271, Tamaqua, PA 18252, U.S.A.

A review of the various methods used for calculating atomic and group electronegativity is given. It is seen that the work falls naturally into two phases or periods. In the first period the emphasis was placed on finding the proper method to calculate atomic electronegativity values. It was assumed that each element can be assigned one value which would describe its electronegativity under all conditions. Many different scales were introduced during this period. Those which are needed to understand current work are reviewed.

The second period concentrated on elucidating the effects of charge and hybridization on electronegativity. The primary concepts behind this work are: the definition of electronegativity as the rate of change of energy of the atom in the molecule with respect to its charge, and the assumption that the atomic electronegativity values become equalized in the molecule. These, along with assumed dependencies of atomic energy on charge, yielded useful methods for calculating both charge distributions in molecules as well as group electronegativity.

Numerical comparisons are shown for both atomic and group electronegativity scales. It is also shown that one of the major concerns common to both periods is the usefulness of the suggested scale or calculational scheme. This is reflected in the tendency for most approaches to be intuitive and easy to use.

1 Introduction 2

2 Classical Scales of Electronegativity 3
2.1 Pauling 3
2.2 Mulliken 4
2.3 Allred-Rochow 5
2.4 Gordy 6
2.5 Sanderson 7
2.6 Other Methods 8
2.7 Atomic Electronegativity – Numerical Comparisons 9
2.8 Summary 10

3 Modern Ideas in Electronegativity 11
3.1 Iczkowski and Margrave 11
3.2 Hinze, Whitehead, and Jaffe 12
3.3 Huheey 15
3.4 Sanderson 17
3.5 Klopman 18
3.6 Other Work 21
3.7 Group Electronegativity – Numerical Comparisons 21
3.8 Summary 22

4 General Overview 23

5 References 24

1 Introduction

Electronegativity is an important part of the intuitive approach to understanding nature that sets chemists off from other physical scientists. Its modern history spans about 50 years. But even to this day there is no definite answer to the question, "What is electronegativity?". This fact leads naturally to the following additional questions: Why is electronegativity so useful to chemists? And why has it had such a long existence?

The reason for the sustained interest appears to lie in the fact that the idea of electronegativity is practically a direct consequence of foundation concepts of modern chemistry, specifically the following three:

1) molecules are made up of atoms held together by chemical bonds.
2) chemical bonds involve a sharing of electrons between the atoms.
3) the electrons are not always shared equally.

Given these three statements then it is almost a matter of human nature to assume that there is something about the atoms that would cause this unequal sharing. That something, of course, is called electronegativity. Hence, the definition of Pauling as the power of an atom in a molecule to attract electrons[1)]. With such close ties to basic concepts it is no wonder that it has been so useful. It appears that at least one conclusion can be drawn from the years of effort: there *is* something about the atoms that causes unequal electron sharing.

Unfortunately, the "something" apparently cannot be defined very precisely. However, because of the potential usefulness of the idea many attempts have been made. Thus, electronegativity has evolved from the initial idea of an assignment of a single value per atom to the present formulation which involves a range of values which depend on the state of the atom in the molecule. As will become apparent this evolution can be broken up naturally into three phases or periods. The first will be dealt with in the first section of this chapter. The work during this period involved for the most part a search for the proper method to calculate or measure atomic electronegativity. The second phase will be dealt with in the second section. During this period the effects of atomic charge on electronegativity were clarified and explored. In addition, the importance of electronegativity equalization was recognized. The emphasis changed to calculation of group electronegativity and atomic charge. A third period has been apparent since the classic paper of Parr[51)]. This phase involves a deepening of the basic concept mainly through the use of density functional theory. This period will not be covered in the present chapter.

The present chapter is concerned with some of the more intuitive methods of arriving at specific values of atomic and group electronegativity that were studied during this evolution. Since the concept is such a basic and important one to chemists there have been many of these suggested over the years. However, since only some of the main ones will be reviewed, the interested reader is referred to several other sources for more detail[2–8)].

2 Classical Scales of Electronegativity

There are five different ways of obtaining specific electronegativity values that can be considered to be classical methods. Practically all of the modern work is related in various degrees to these efforts. These are the scales of Pauling[1, 9], Mulliken[10], Sanderson[11], Allred-Rochow[12], and Gordy[13], respectively. Taken together they give a good idea of the range of meaning that can be attributed to the concept. Three of these, i.e. Allred-Rochow, Gordy and Mulliken, can be seen as attempts to answer the question, "What is electronegativity?". The other two are more properly considered to be primarily correlations of empirical data to electronegativity.

One of the basic problems being addressed by all of these authors is to provide a set of numbers or a way to get a set of numbers which can predict or explain charge distributions in molecules. The development of these methods took place between 1932 and 1958.

2.1 Pauling

All new scales of electronegativity or new methods of calculating values of electronegativity are referred to Pauling's original scale. If electronegativity were a measurable quantity its units would probably be in terms of "Paulings". The reasons for this appear to be two-fold, i.e. this scale was published first and for years it had more specific values available than any other scale. However, in terms of present understanding it should probably be considered one of several available useful empirical correlations rather than as a preferred means of understanding the fundamental concept.

The original ideas were introduced in 1932[9]. These in turn were predicated by results of work done previously which concerned the nature of the chemical bond[14]. Basically, the earlier work suggested that it is reasonable to treat bonds as being independent entities. One example of this is with regard to bond energies, i.e. there was evidence to support the postulate of the additivity of normal oovalent bonds as given by the relation

$$D^\circ(AB) = (D(AA) + D(BB))/2 \qquad (1)$$

where $D^\circ(AB)$ is the energy of a normal covalent bond between atoms A and B.

In addition, straightforward quantum mechanical arguments indicated that the presence of ionic character (i.e., unequal electron sharing) in the bond would increase the energy over its covalent value. That means that the actual D(AB) would be greater than $D^\circ(AB)$ from (1). This was found to be true experimentally. If Δ is the difference between these two energies than Pauling's famous electronegativity postulate is given by

$$\Delta = (\chi_A - \chi_B)^2 \qquad (2)$$

where χ_A and χ_B are the electronegativities of atoms A and B. It should be noted that electronegativity difference is related to $\sqrt{\Delta}$ (rather than Δ) since Pauling found that these and not the Δ were additive. Pauling's scale was based on this postulate. Because of the availability of thermochemical data he was able to assign values to a large number of atoms.

It was soon discovered that Δ was not always positive[15]. This along with other theoretical considerations led Pauling to modify (1) to give

$$\Delta = D(AB) - \sqrt{D(AA)D(BB)} \tag{3}$$

Arbitrarily setting the value of H at 2.1 yields:

$$|\chi_A - \chi_B| = 0.208\sqrt{\Delta} \tag{4}$$

Most often, Eq. (1) is used to arrive at a value for Δ. Several retabulations and extensions of Pauling's original work have been done using more recent data[16, 17].

The empirical usefulness of this scale has been confirmed in many investigations[2]. However, as will be discussed later, the use of a single value for each atom is not sufficient to treat all cases[18, 19]. It appears to be generally accepted that theoretical confirmation of Pauling's method of obtaining his scale is weak[7]. Because of the fact that one number is assigned to each atom, changes of hybridization are not readily accounted for. In addition, effects of atomic charges are completely neglected[20].

2.2 *Mulliken*

Pauling's scale was the first of the modern attempts at quantifying electronegativity. The second was due to Mulliken[10]. His measure is simply the average of the first ionization potential (I_A) and the electron affinity (A_A), i.e.

$$\chi_A^M = (I_A + A_A)/2 \tag{5}$$

Although Mulliken's method has much greater theoretical support[7, 21, 22], Pauling's remains the accepted scale of reference. This is primarily because up until the early 1960's there was insufficient data available to provide Mulliken electronegativity values for many atoms. Even today there is little data available to cover the vast majority of interesting cases among the transition elements. Thus, this measure of electronegativity is of little use even now with this important group of atoms.

However, there appears to be much support in favor of the idea that this scale gives the best representation of the electronegativity concept[7]. This is not in terms of its quantitative usefulness so much as its theoretical content. The vast majority of current theoretical work in electronegativity can be viewed either as derived from or highly dependent on Mulliken's ideas.

Aside from its theoretical support the scale appears reasonable even on an intuitive level[23]. A bond can be seen as a competition between two atoms for a pair of electrons. Electronegativity would then be expected to represent the relative ability of each atom to compete for the electrons. Each atom attempts to keep one electron (i.e. resist becoming a positive ion) and simultaneously acquire the second electron (i.e. become a negative ion). These two processes can be seen as involving the ionization potential and electron affinity respectively. Thus, it is not unreasonable to use an average of these two quantities to give a measure of the competition and, therefore, the electronegativity.

I_A and A_A values can be calculated for atoms in any state desired (ground, excited, valence). Thus, the method is quite versatile in being able to represent the atom as it exists in the molecule. In addition, it can be rendered charge-dependent simply by obtaining I_A and A_A values for the charged species. A series of papers appearing in the early 1960's provided chemists with an extensive tabulation of Mulliken values for non-transition series atoms in various valence states[24, 25]. Unfortunately, electronegativity values were obtained only for members of the first transition series. This still remains a serious drawback. The calculated values are linearly related to Pauling's via the following equations

$$\chi_A = 0.168\,(\chi_A^M - 1.23) \tag{6}$$

In this same series of papers, Hinze, Whitehead, and Jaffe stressed the concept of orbital electronegativity[25]. This refers to the electronegativity of the orbital which the atom will use in its bond. This, of course, implies a point of view requiring the use of atomic orbitals in molecules. It does, however, clarify the reason for the different electronegativity values which, for example, C might exhibit in different circumstances. Thus, C in acetylene has a higher electronegativity than C in methane because it uses an sp hybrid orbital in bonding rather than an sp^3 hybrid.

Not only does the orbital concept explain different behaviors but it also adds precision to the "atom in the molecule" restriction. According to the understanding of quantum mechanics at the time these papers were published, the idea of an atom in a molecule was devoid of meaning in any reasonably normal sense. Thus, the attraction of this "atom" for electrons would be even less meaningful. This would tend to negate the whole electronegativity concept. However, an atomic orbital is well defined mathematically and readily understood intuitively. Thus, using the orbital concept and, therefore the orbital electronegativity extension, provides precision and meaning as well as preserving the atom in a molecule point of view.

One of the results of the Hinze-Jaffe calculations is a clear recognition of the effect of atomic valence state and orbital hybridization on electronegativity values. In general, for sp hybridized orbitals greater s character implies greater electronegativity. It was in fact found that the dependence is linear. For spd hybrids the relation is more complicated.

2.3 Allred-Rochow

Aside from Pauling's, the Allred-Rochow scale is probably the most often used as a measure of electronegativity[12]. It is based on the simple assumption that the electronegativity of an atom is given by the force of attraction between the screened nucleus and an electron at the covalent radius. This is expressed as

$$\chi_A = 0.36\,Z_{eff}/r^2 + 0.74 \tag{7}$$

Where Z_{eff} is an effective nuclear charge obtained from Slater's rules, r is the covalent radius and the coefficients put the scale into Pauling units. As is implied by (7) this scale is linearly related to Pauling's.

This work is significant for several reasons. First, it introduces the idea of force into electronegativity theory. This appears to be quite consistent with Pauling's verbal definition and also easy to understand and modify. Secondly, it exemplifies the importance of the idea of maintaining calculational simplicity to achieve usefulness. Both Z_{eff} and r are readily available quantities for many elements. This fact, coupled with the simplicity of Eq. (7) has made this scale one of the most heavily used. In addition, it opened up the possibility of studying more elements than was previously possible.

Several recent modifications and extensions of these basic ideas are worth noting. The first is due to Huheey[26]. He makes two assumptions, i.e. that radius is inversely proportional to effective nuclear charge (Z_{eff}) and that Z_{eff} varies linearly with partial atomic charge (δ). These lead to the following approximate expression

$$\chi_A = 0.36\,(Z_{eff} - 3\,\delta)/r^2 + 0.74 \quad (8)$$

Although the method is admittedly simplistic in nature, Huheey shows[26] that it leads to results which are similar to those obtained in some of the more rigorous methods described below.

Recently Boyd and Markus have performed nonempirical calculations utilizing the force concept[27]. In their work χ_A is given by

$$\chi_A = \frac{kZ_A}{r_A^2}\left[1 - \int_0^{r_A} D(r)dr\right] \quad (9)$$

where Z_A and r_A are atomic number and relative radius of A and D(r) is the radial density function. Setting $\chi_F = 4$ gives $k = 69.4793$. r_A was chosen in terms of the electron density contour corresponding to 10^{-4} atomic units for each atom. Using Hartree Fock wave functions, the authors obtained good correlation with the other classical scales.

A third extension of the Allred-Rochow idea is due to Mande et al.[28]. They use an effective nuclear charge obtained from X-ray spectroscopic data. Since they use experimental data the values used for effective nuclear charge are less arbitrary than Slater's and, thus, the electronegativity scale should be more reliable. They also obtain a good correlation with Pauling's scale. In addition, the results they obtain for the first transition series elements appear more reasonable than Allred-Rochow's.

In addition to these papers, a recent effort by Zhang has led to a modification of the method to account for oxidation states of the atoms[29]. All of this work underlines the basic usefulness and versatility of the force concept in electronegativity theory.

2.4 *Gordy*

Gordy has suggested a number of ways of arriving at electronegativity values[13, 30, 31]. Only one will be treated here[13] since the other two fall into the category of correlations with empirical data and do not directly address the question as to the meaning of electronegativity. Gordy considered electronegativity to be the electrostatic potential at the covalent radius caused by the screened nuclear charge. Specifically, this is

$$\chi_A = 0.62\,(Z'/r) + 0.50 \quad (10)$$

Where Z′ is a screened charge obtained using Gordy's method, i.e. closed-shell electrons screen totally (i.e. screening factor of 1) and valence electrons have a screening factor of 0.5. Thus, if the atom has V valence electrons $Z' = V - 0.5\,(V - 1) = 0.5\,(V + 1)$. Relation (8) becomes

$$\chi_A = 0.31\,(V + 1)/r + 0.5 \tag{11}$$

Note that the constant terms in (10) and (11) put this scale into Pauling units. If a more reasonable method is used for Z′ in (10), e.g. Slater's values, the correlation with Pauling values is not as good[2)].

This scale is not used very often by investigators other than Gordy. However, it is important for several reasons. First, it introduces the idea of a potential into electronegativity theory. Secondly, as with the Allred-Rochow method, it is simple to use and, thus, should be quite useful. However, because of the difficulties involved in the estimation of screened nuclear charge, this has not been the case. Recently, Parr and Politzer have found that there is some merit to using this definition with a different screened nuclear charge[32)].

The comparison between the Allred-Rochow and Gordy scales is interesting. On the one hand, they are quite similar in form, while on the other, they represent two different physical parameters. Reasonable arguments can be presented for either as suitable measures of electronegativity. However, conformity with experiment should decide the issue. Unfortunately, it does not. The best that can be said at present is that the Allred-Rochow scale appears to have been correlated with the bulk of the data. Recent thinking in electronegativity theory, however, leans more toward the potential point of view[33)]. This would tend to give Gordy's scale more theoretical support. Thus, even though it would appear to be a simple choice between the two methods, the vagueness of the electronegativity concept itself precludes deciding.

2.5 *Sanderson*

This scale has also not been used by investigators other than its originator. Recently, however, it has been getting more attention[5, 34–36)]. Probably the main reason for its not being used is its apparent obscurity. The method is based on the idea of a stability ratio[11)]. This is simply the ratio of the average electron density of the atom (ED) to the electron density of an isoelectronic inert atom (ED°). ED is given by

$$ED = Z/4.19\,r^3 \tag{12}$$

in which Z is the number of electrons on the atom and r is the covalent radius. In the case of the isoelectronic inert atoms, r is an interpolated value and, thus, is fictional. The stability ratio (SR), thus obtained, is a measure of the compactness of the atom, i.e. how tightly the electrons are held. Sanderson claims that this is related to electronegativity. His argument goes as follows[37)]. SR gives a measure of how well the atom is able to hold onto its own electrons. The more compact, the greater the holding power. He argues that if atom A holds its electrons tighter than atom B, then it will also do a better job of attracting bonding electrons in a molecule. In fact, SR can be related to Pauling's values as follows

$$\chi_A = (0.21\,(SR) + 0.77)^2 \quad (13)$$

Sanderson has used these values along with an electronegativity equalization scheme and other assumptions to derive thermochemical bond data for many groups[37)]. The results obtained appear to correlate well with experimental data. However, this method, like the associated electronegativity scale, has not been extensively used by others. Perhaps a clearer connection with other work would change this situation.

2.6 Other Methods

Three other scales of atomic electronegativity have been suggested since the classical period, i.e. after 1958, which should be mentioned.

The first is a purely theoretical scale[38)]. It represents the first attempt to derive atomic electronegativity values without using empirical data. The method obtains electronegativity from floating spherical Gaussian orbital (FSGO) wavefunctions[39)]. Bonds are described by Gaussian orbitals which are allowed to float to a location of minimum energy between the atoms. The authors define an orbital multiplier f_{AB} as

$$f_{AB} = R_A/(R_A + R_B) \quad (14)$$

where R_A and R_B are distances from the atoms A and B to the orbital center. If $f_{AB} = 0.5$ both A and B attract equally. If $f_{AB} < 0.5$ then A attracts greater than B. The simplest definition of electronegativity is in terms of the difference

$$\chi_A - \chi_B = K\,(f_{AB} - 0.5) \quad (15)$$

(15) is used with $\chi_{Li} = 1$ and $\chi_F = 4.0$ to establish the scale. The results are shown to be quite consistent with Pauling and Allred-Rochow's scales.

The second scale was developed by St. John and Bloch[40)]. It is based on the Pauli-force model potential[41)]. They define an orbital electronegativity (X_l) for valence orbitals as

$$X_l = 1/r_l \quad (16)$$

in which r_l is a radius for the valence orbital with quantum number l derived from the Pauli model potential fitted to experimental data. Note that X_0, X_1, X_2 correspond to s, p, and d orbital electronegativity respectively. Atomic electronegativity is given by

$$\chi_A = 0.43 \sum_{l=0}^{2} X_l + 0.24 \quad (17)$$

As with previous scales, the constants put this scale into Pauling units. This scale, as Gordy's, is related to the electrostatic potential idea. Unlike Gordy's, however, it introduces the idea of hybridization in an explicit manner.

Both of these scales represent the kind of work that characterizes this first period. However, they also indicate the extensions that are possible due to advances in other fields.

An empirical optical electronegativity scale, χ_{op}, was proposed by Jørgensen[42a)] with the purpose of rationalizing the electron transfer spectra of transition metal complexes, MX. A linear difference in χ_{op} was sought to represent the photon energy hv of the first Laporte allowed electron transfer band in a metal-ligand system according to the relation $h\nu = [\chi_{op}(X) - \chi_{op}(M)] \cdot 30{,}000\ cm^{-1}$. The choice of the constant $= 30{,}000\ cm^{-1} = 3.7\ eV$ normalizes the χ_{op} values of halogens to the Pauling electronegativity. For the highest occupied M.O. of halides the ionization energy given by (3.7 eV) χ_{op} is within 0–1 eV for nearly all gaseous halides whereas for solids, this difference is within -1 to 0 eV[42b)]. A linear relationship of χ_{op} to the difference in eigenvalues as realized by Jørgensen is an important idea which can be rationalized in terms of the density functional theoretic approach to χ.

2.7 *Atomic Electronegativity – Numerical Comparisons*

Table 1 presents a comparison of the five classical methods for some common elements. A correlation coefficient has been calculated for four of these versus Pauling's scale. It can be seen that the correlation is good in all cases with the worst obtained for the Mulliken-Jaffe values. This is probably due to the uncertainty involved in choosing the

Table 1. Comparison of selected electronegativity values[a]

Atom	Pauling	Mulliken-Jaffe (Orbital)	Allred-Rochow	Gordy	Sanderson
H	2.2	2.21 (s)	2.20	2.17	2.31
Li	0.98	0.84 (s)	0.97	0.96	0.86
Be	1.57	1.40 (sp)	1.47	1.38	1.61
B	2.04	1.93 (sp^2)	2.01	1.91	1.88
C	2.55	2.48 (sp^3)	2.50	2.52	2.47
N	3.04	2.28 (p)	3.07	3.01	2.93
O	3.44	3.04 (p)	3.50	3.47	3.46
F	3.98	3.90 (p)	4.10	3.94	3.92
Na	0.93	0.74 (s)	1.01	0.90	0.85
Si	1.90	2.25 (sp^3)	1.74	1.82	1.74
Cl	3.16	2.95 (p)	2.83	3.00	3.28
Ge	2.01	2.50 (sp^3)	2.02	1.77	2.31
Br	2.96	2.62 (p)	2.74	2.68	2.96
Sn(IV)	1.96	2.44 (sp^3)	1.72	–	2.02
I	2.66	2.52 (p)	2.21	2.36	2.50
Correlation coefficient	–	0.926	0.983	0.993	0.989

[a] All from Ref. 5 except Gordy which is from: Gordy, W., Cook, R. L.: Microwave Molecular Spectra: In: Technique of Organic Chemistry 9, 2, 2nd E. (West, W., Ed.), Wiley (Interscience) 1970

orbital hybridization. It is clear from this comparison that the scales are reasonably consistent with one another. However, as Huheey points out, values are probably not interchangeable among scales[5)].

2.8 Summary

Although not all work done during this period is represented here, it is believed that the general flavor of the theoretical efforts can be gleaned from these five scales. Several general observations can be made with regard to these scales which should have universal validity.

(1) It can be seen that all methods are conceptually simple. They even border somewhat on the intuitive. In addition, the more successful ones (Allred-Rochow and Pauling) make it easy for the investigator by providing data for many atoms. Thus, it appears that two of the early hallmarks of the electronegativity concept are conceptual simplicity and ease of use. As will become apparent in the next section, both of these characteristics have been carried over into the second phase of electronegativity theory development.

(2) None of the scales have the same units. The units can be summarized as follows

1) Pauling – $(\text{Energy})^{1/2}$
2) Mulliken – Energy
3) Allred-Rochow – Force
4) Gordy – Energy/electron
5) Sanderson – Dimensionless

Thus, it is obvious that there is no agreement as to the specific physical meaning of electronegativity. It has even been suggested that there is no meaning and thus the concept should be rejected[43)]. The proven usefulness of the concept and the lack of easily used and inexpensive rigorous quantum mechanical methods makes this position unreasonable.

Since about 1961, the tendency has been toward an acceptance of electrostatic potential (energy/electron) as the measure of electronegativity[33)]. The situation, however, is still not totally clear. Whether the question of units (and, therefore, the specific physical content) is ever resolved, the concept will still be useful. This is because it is on the same conceptual plane as the idea of an "atom in a molecule" and "chemical bond". When these latter concepts are no longer needed, then electronegativity will also become unnecessary.

(3) Mulliken's scale is probably the most correct since it is the soundest theoretically[7)], but it suffers from the need to specify the hybridization of the orbitals used. The other scales are easier to use when hybridization is not known since they, in a sense, already account for hybrid effects in the empirical parameters used. Huheey had pointed this out several years ago[5)]. He further suggested that different scales not be mixed. By properly choosing different values from different scales practically any desired result could be obtained. His advice appears to be sound, i.e. if possible choose Mulliken values; if not, then any of the others are probably equally as good as each other.

(4) Although there is some effort to take other effects into account, the main thrust is to provide one number per element to represent its electronegativity. Thus, none of the

scales treat charge explicitly and only Mulliken's handles various hybrid orbitals in a natural way. It is true that during the period of development of these methods it was recognized that other effects should be taken into account[12]. It can also be seen that there was an evolution from Pauling's concept of a single number per element to the idea of a range of values per element. However, it was left to the investigators who came after these pioneers to clarify and quantify these ideas.

3 Modern Ideas in Electronegativity

Most of the theoretical work in electronegativity in the past 20 years has centered around two main ideas, viz. electronegativity equalization[44] and the identification of electronegativity as the rate of change of energy per change in atom charge[33]. Of course, more work has been done in studying both concepts than is reported here[45–54]. These other efforts, however, have not been of particular use in providing improved electronegativity calculational schemes. Thus, they will not be treated. The culmination of this work appears to be the recent efforts of Parr and collaborators which has given theoretical support to both of these concepts[55].

The combining of these two concepts has led to a deeper understanding of electronegativity. In addition, it has resulted in quite useful methods for calculating both atomic charge in a molecule and group electronegativity[25, 33, 56–58].

3.1 Iczkowski and Margrave

If any work heralded the new direction in electronegativity theory, it would be the paper of Iczkowski and Margrave[33]. They first noted that the energy of an atom $A(E_A)$ can be expressed as a function of the charge (δ) on the atom, i.e.

$$E_A(\delta) = a\delta + b\delta^2 + c\delta^3 + d\delta^4 \tag{18}$$

in which a, b, c, and d are constants that depend on the atom and the valence state it is in. They then argued that the electronegativity of atom A should be given by

$$\chi_A = -(dE_A/d\delta)_{\delta=0} \tag{19}$$

As support for this assumption, they noted that χ_A so defined reduces exactly to Mulliken's definition if only the first two terms are kept in (18), i.e. if E_A is given by

$$E_A(\delta) = a\delta + b\delta^2 \tag{20}$$

The data presented in their paper shows that to a very good approximation this is true. Thus, the definition expressed by Eq. (19) appears quite resonable. With this approximate energy relationship, electronegativity becomes

$$\chi_A = a + 2b\delta \tag{21}$$

Equations such as (19) and (21) represent the current thinking in electronegativity theory. Note that they follow the past trend with[2] regard to simplicity of concept and, as will be seen, ease of use.

It is readily apparent that this work represents a radical departure from the classical scales presented above. More than just a formula for calculating specific values, it provides a new approach to understanding electronegativity. Since it reduces to Mulliken's scale as an approximation it also gives an idea as to the level of approximation at which all of the earlier scales are useful. Also, since the concept is expressed in general terms, i.e. a derivative of energy with respect to charge, it is easily amenable to further generalization[55] and can also be used with higher level approximations to the energy of the atom[59].

Several other points should be noted with regard to this paper. First, it introduces an explicit dependence on atom charge. Although this is not the first reference to the importance of charge[1, 44] it is certainly the first time that it had been treated so naturally and explicitly. Secondly, the work relies for its authority on Mulliken's scale. Most of the current theories of this type follow the same pattern[25, 55]. Thus, it is being tacitly assumed that Mulliken's work is in some sense a good measure of electronegativity or possibly even the best measure of electronegativity. As mentioned above, the Mulliken scale makes intuitive sense and also has theoretical support. However, the derivation does involve assumptions which some believe to be severe[7]. In addition, because of the vagueness of the basic concept, it probably can never be certain that any one scheme is the correct measure of electronegativity. Thus, in a sense all of the modern work rests on somewhat shaky foundations. Each effort must then rely on its own usefulness and ability to correlate experimental data for its justification. As we will see this is being done.

3.2 Hinze, Whitehead, and Jaffe

A series of papers appeared within a year of Iczkowski and Margraves which were also quite important in the development of electronegativity theory[24, 25]. These were the papers of Hinze, Whitehead, and Jaffe. As mentioned above, they provided the first extensive set of calculations of electronegativity values based on Mulliken's definition and they also introduced the important idea of orbital electronegativity. In addition, they extended Iczkowski and Margrave's use of the idea of a potential to include atomic orbitals. To do this, they defined electronegativity (now orbital electronegativity) as the derivative of the energy of an atomic orbital j with respect to the occupation (n_j) of the orbital,

$$\chi_{A,j} = \partial E_A / \partial n_j \qquad (22)$$

They observed that atomic electronegativity is most reasonably taken as referring to the atomic orbital before bonding. This translates to the case of a half filled orbital, i.e. $n_j = 1$. If, in addition, the energy of the orbital is assumed to be a quadratic function of n_j, then their definition of atomic electronegativity reduces to Mulliken's. This fact is taken as adequate justification for making the assumptions. In addition, with these assumptions electronegativity can now also be defined for the cases in which $n_j = 0$ and 2 i.e. for both empty orbitals and lone pairs. These are useful in situations involving electron pair donor/acceptor bonds. But it can be seen that these are the only three cases in which the concept of electronegativity appears valid, i.e. for $n_J = 0, 1$, and 2. However, with the definition given in (22) it also becomes possible to define a term which is valid for non-integral values of n_J called "bond electronegativity". To do this, consider a bond to be formed in two steps, i.e. electron pairing followed by electron transfer between the atoms. Transfer of charge from atom A to atom B requires energy changes of $(\partial E_A/\partial n_A)$ dn_A and $(\partial E_B/\partial n_B)$ dn_B respectively. At equilibrium there is no further change in energy. Since $dn_A = dn_B$, then

$$\partial E_B/\partial n_B = \partial E_A/\partial n_A \tag{23}$$

i.e. the electronegativities are equalized on bond formation. The electronegativity value attained by the atom in bond formation is called "bond electronegativity". The authors point out that this should not be confused with the Pauling electronegativity which is appropriate only for integral values of orbital occupation as outlined above.

It should be noted that the idea of electronegativity equalization which is so important to current theory did not originate in these papers. It is Sanderson who first postulated that the electronegativity of all atoms in a molecule or group should become equalized[44]. These authors placed it on a firmer foundation and related it in a clear fashion to other electronegativity methods, in particular Mulliken's.

Note that Eqs. (22) and (23) along with the assumed quadratic dependence of energy on n_J gives (in the notation of Hinze et al.)

$$b_A + 2\,c_A\,n_{A,J} = b_B + 2\,c_B\,n_{B,J} \tag{24}$$

where the b and c terms are constants depending on the valence state and orbital being considered. Since $n_{A,J} + n_{B,J} = 2$ (i.e. 2 electrons per bond), Eq. (24) can be used to estimate the ionic character of the bond as well as orbital (and, therefore, atomic) charge.

Note further that the definition of orbital electronegativity given above also allows the calculation of group electronegativity. This is and has been of great importance especially in organic chemistry[4]. It is well known that the electronegativity of $-CH_3$ is not the same as $-CCl_3$ or $-CCH$ even though the three groups bond through the C atom. This is because electronegativity depends on hybridization and charge on the central atom. Both of these facts are expressed in the equation for electronegativity, i.e.

$$\chi_{A,J} = b_A + 2\,c_A\,n_{A,J} \tag{25}$$

simply by noting that both b_A and c_A are functions of hybridization and charge. Hinze gives a fairly complicated graphical method to arrive at the specific dependence on charge. Using this information plus a self-consistent iteration scheme, it is possible to arrive at charge distributions in the group. This allows the calculation of group electronegativity values. They compared values obtained by this method with other data available at the time with good results[25].

As can be seen, this series of papers introduced several important concepts and methods into the electronegativity theory arsenal. Accordingly, it has made an impact on most subsequent work in the field. As with any effort of this type, there are areas that can be criticized. One of these is the use of the electronegativity equalization principle. It has been argued that this approach is somewhat simplistic since it neglects resonance and electrostatic effects[56, 60–62]. Results obtained by Pritchard indicate that the electronegativity remains unequal by an amount which is of the order of 10% of the original difference[60]. More recent work, however, suggests that equalization does occur[49, 55]. This will be discussed further below. A second area of criticism involves the heavy emphasis placed upon the orbital concept. Although this is definitely a useful idea, it is probably not justified to claim that electronegativity refers only to orbitals. As other work has shown, it is not even necessary to use the idea of orbitals to discuss electronegativity[50]. Other authors also suggest that recourse to the orbital concept does not solve the problem of assigning meaning to the "atom in a molecule" portion of the electronegativity concept[7].

Because one of the desirable elements of an electronegativity calculational scheme is simplicity and/or ease of use, all of these methods must be judged by this standard. The fact that the orbital viewpoint in conjunction with electronegativity equalization allows the calculation of group electronegativity is significant. However, the method proposed by these authors offers little incentive over other theoretical or semi-empirical methods.

Because of the potential usefulness of group electronegativities, several authors have investigated the possibility of devising simple schemes to perform the calculations[58, 59]. The first to be discussed here is a paper by Whitehead, Baird, and Kaplansky[63]. In order to understand their contribution, it must first be noted that b_A and c_A in (25) can be written in terms of ionization potential and electron affinity

$$b_A = (3\,I_{a,j} - A_{A,j})/2 \tag{26}$$

$$c_A = (A_{A,j} - I_{A,j})/2 \tag{27}$$

in which specific reference to the orbital j is inluded for completeness. If n_T is the total occupation of all of the valence orbitals in A except orbital j, then Whitehead showed that $I_{A,j}$ and $A_{a,j}$ can be approximated as

$$I_{A,j} = \propto_j + \beta_j n_T + \gamma_j n_T^2 \tag{28}$$

$$A_{A,j} = \delta_j + \varepsilon_j n_T + \delta_j n_T^2 \tag{29}$$

where α_j, β_j, etc. are constants which depend on the hybridization of the valence orbitals other than j.

Equations (28) and (29) can be substituted into (26) and (27) to give b_A and c_A in terms of n_T. These can then be used in (25) to express the orbital electronegativity as a function of n_T. The authors then assumed orbital electronegativity equalization and that $n_A + n_B = 2$ for each bond to calculate charge distribution. Since the final equations are non-linear, an iteration procedure was necessary.

It was shown that the values obtained in this way were essentially the same as those obtained using the procedure suggested by Hinze et al. However, the labor involved was much less. The authors compared results to NMR and NQR data and obtained a good correlation[63]. This same method was compared to the Extended Hückel Theory and additional NMR chemical shift data[64]. It was shown that the values derived from the electronegativity method exhibited greater correlation, even though the calculational effort is similar in both cases. In particular, it was shown that the method predicts the experimental inductive effect while the EHT method does not.

Even though this work represents an improvement over the Hinze method with regard to simplicity, it still apparently did not totally meet the implied criterion for electronegativity theories. However, it did show that theories of this type can be quite useful and, thus, worthy of more effort.

3.3 Huheey

In 1965, Huheey presented a much simpler procedure for calculating group electronegativity[56]. He took Iczkowski and Margrave's Eq. (21) as his starting point. But now the a and b are derived from Hinze and Jaffe's values for ionization potential and electron affinities[24]. Thus, his electronegativity refers to an atomic orbital. In addition, he used an electronegativity equalization principle. But rather than the one proposed by Hinze, Jaffe and Whitehead, he chose Sanderson's[44]. As mentioned above, this assumes complete equalization of the electronegativity of all of the atoms (in this case all atomic orbitals involved in bonding) to one average value.

In this method, for a group $-AB_n$, the electronegativity of each atom is set equal to each other atom, i.e.

$$\chi_G = \chi_A(\delta_A) = a_A + 2\,b_A\delta_A = a_{B_1} + 2\,b_{B_1}\delta_{B_1} = \ldots = \chi_{B_n}(\delta_{B_n}) \tag{30}$$

where the dependence of χ_A on δ_A is made explicit for clarity. This yields n equations in n + 1 unknowns (δ_A, ---, δ_{B_n}). These equations are then coupled separately with each of the following three relations

$$\delta_A + \delta_{B_1} + \ldots + \delta_{B_n} = 0 \tag{31}$$

$$\delta_A + \delta_{B_1} + \ldots + \delta_{B_n} = 1 \tag{32}$$

$$\delta_A + \delta_{B_1} + \ldots + \delta_{B_n} = -1 \tag{33}$$

to yield charge distributions and group electronegativity values for the group as radical ($\delta_G = 0$), anion ($\delta_G = 1$) and cation ($\delta_G = -1$) respectively. This gives a straight line when group charge (δ_G) is plotted versus group electronegativity ($\chi_G(\delta_G)$). Denoting the slope of the line as b_G and the charge intercept as a_G gives

$$\chi_G = a_G + b_G\delta_G \tag{34}$$

which relates group electronegativity to charge. Note that a_G is the normal group electronegativity (called inherent electronegativity by Huheey).

Huheey calculated the electronegativity values for 99 different groups using this scheme. He showed these values to be reasonably consistent with experimental values and very similar to the values obtained by Hinze using the longer procedure.

As it stands, Huheey's procedure is not able to treat groups with multiple bonds. This is because in general a and b associated with σ and π orbitals are different. Thus, for atom A involved in a multiple bond, electronegativity equalization requires that

$$a_{A\sigma} + b_{A\sigma}\delta_A = a_{A\pi} + b_{A\pi}\delta_A \quad (35)$$

for A. This is impossible for $a_{A\sigma} \neq a_{A\pi}$ and $b_{A\sigma} \neq b_{A\pi}$. Huheey circumvented this problem by using a procedure which essentially averages the results obtained using $a_{A\sigma}$, $b_{A\sigma}$ with those obtained using $a_{A\pi}$, $b_{A\pi}$. In this way he was able to calculate the electronegativities of an additional 97 groups.

This work represents the first time that such a large set of group electronegativity values was obtained from a theoretical scheme requiring only atomic parameters. The values have been used in several different types of investigations. These range from research regarding group electronegativity to correlations with experimental data[4, 65].

Huheey also pointed out the importance of the charge coefficient term (b_A or b_G) in both group and atomic (or orbital) electronegativity. Thus, a low value of b_G tends to buffer the electronegativity of the group toward charges. This is because it takes a larger charge shift to move the group electronegativity either up or down if the b term is small. In Huheey's work a larger group resulted in a smaller value of b. This suggests, for example, that the larger an alkyl group is the more charge it can absorb per change in electronegativity of the group. He showed this to be valuable, e.g. in explaining acidities of various aliphatic alcohols[65]. He further points out a relationship between the b term and the hardness and the softness of an atom, and presumably of a group. Since b is the second derivative of energy with regard to charge this anticipates and is consistent with Parr and Pearson's recent work in this area[66].

The method developed by Huheey appears to have three major drawbacks: it is not able to account for differences in isomers (e.g. $-CH_2CFH_2$ and $-CFHCH_3$), it is not able to treat groups with multiple bonding easily and it apparently overestimates the effect of the atoms or groups attached to the bonding atom (e.g. the effects of the H atoms in NH_2)[4]. The method has also been criticized by Huheey himself in that it relies on complete electronegativity equalization[67]. In fact, he has shown that a modification of the scheme to provide for only 80% equalization resolves the first problem[67]. It appears, however, that Huheey continues to use the complete rather that the partial electronegativity equalization concept in his work[5, 68].

Recently a very simple scheme was presented to account for the isomer problem[69]. It is based on the assumption that inductive effects are not operative beyond three bonds. In addition, it assumes that the b value for the group is the same for all isomers. Based on these assumptions, it uses a weighting scheme for the a term (inherent electronegativity) which weighs the a value of the central atom directly attached to the group of interest most heavily, the next nearest central atom less, etc. To illustrate, consider the group

```
      CH2CH2Cl
      |
-CH2-CHCH2Cl
```

In this case a_G is given by

$$a_G = \frac{3^3 a_{CH_2} + 3^2 a_{CH} + 3^1 a_{CH_2Cl} + 3^1 a_{CH_2} + 3^0 a_{CH_2Cl}}{3^3 + 3^2 + 3^1 + 3^1 + 3^0} \quad (36)$$

Using this procedure the authors were able to correlate electronegativity and regiochemistry in asymmetric sulfide chlorinations for fifty known instances[69].

It should be mentioned that Huheey's method, as Hinze's, allows the calculation of both group electronegativity and atomic charges. This latter quantity is, of course, related to the unequal sharing of electrons in a bond and represents one of the main reasons for the very existence of the electronegativity concept. Atomic charges calculated with this method have been compared to isomer shifts as well as theoretical values with reasonable results[69, 70]. However, more work needs to be done in this area. Since electronegativity schemes such as Huheey's can provide rapid methods to obtain atomic charges, they have received some attention lately[36, 53, 57] and probably will be used more in the future.

In 1984, a further simplification of Huheey's method was presented[57]. This uses Sanderson's electronegativity equalization principle as well as an observation by Huheey on the approximate behavior of electronegativity with charge. The following very simple relations are derived

$$\chi = \chi_A (1 + \delta_A) \tag{37}$$

$$\delta_A = (\chi/\chi_A) - 1 \tag{38}$$

$$\chi = (N + \delta_G)/\Sigma (\nu/\chi_A) \tag{39}$$

where χ is the equalized electronegativity for the group or molecule, ν is the number of A atoms in the molecule, N (= $\Sigma\nu$) is the total number of atoms, and δ_G is again the charge in the group. For $\delta_G = 0$ Eq. (39) gives the group electronegativity.

This method represents the ultimate in simplicity within the Huheey scheme. The major assumption being made to allow the simplification is the modification of the atomic electronegativity equation from

$$\chi_A = a_A + b_A\delta_A \tag{40}$$

to

$$\chi_A = a_A (1 + \delta_A) \tag{41}$$

This, of course, involves the identification of the charge coefficient b_A with the inherent electronegativity a_A, i.e. it is assumed that $b_A = a_A$. In other recent work it has been shown, however, that a more reasonable estimate is $b_A = 1.5\ a_A$[58]. This will be discussed further below. Of course, since the equations are expressed in terms of χ_A, any method of arriving at inherent or atomic electronegativity can be used (i.e. not only Mulliken's). The author has compared the group electronegativity values obtained using Pauling's estimates of χ_A to other methods with good results.

As with Huheey's, or indeed any method relying on total electronegativity equalization, this method suffers from the three deficiencies mentioned above. However, in line with the basic criterion of simplicity, it represents an advance over previous methods.

3.4 Sanderson

Sanderson also recognized that electronegativity is a function of atomic charge[37]. However, his method for calculating group electronegativity and charge distribution in molecules does not logically fit within the framework established by the other methods except in its use of an electronegativity equalization principle[46]. As mentioned above, this concept was first promulgated by Sanderson. He argued that in the formation of a molecule (or radical) each of the originally different atomic electronegativity values become equalized to one value. He uses the geometric mean of all of the atomic electronegativity values as the final value. In addition to this simple method for arriving at electronegativity an equally simple scheme was used to estimate atomic charge. This is defined as "the ratio of the electronegativity change undergone in forming the compound to the change that would correspond to the acquisition of unit (+ or −) charge"[37]. The unit change in electronegativity (ΔSR) is obtained from the original electronegativity (SR) using the relation

$$\Delta SR = 2.08\sqrt{SR} \tag{42}$$

Thus, partial charge on F in HF is obtained as

$$\delta_F = -\left(\frac{5.75 - \sqrt{5.75 \times 3.55}}{2.08\sqrt{5.75}}\right) = -0.25 \tag{43}$$

where 5.75 and 3.55 are the atomic electronegativity values of F and H.

Because of the simplicity of obtaining both atomic charges and group electronegativities from these procedures many scales are being compared to Sanderson's[36, 57]. However, his procedure suffers from the same three problems as Huheey's.

3.5 *Klopman*

Klopman provided theoretical support and clarification for the methods proposed by Iczkowski and Margrave and by Hinze, Whitehead, and Jaffe[71]. First, he noted that the total energy for a free atom in a particular configuration can be given by

$$E_A = \sum_i B_A^l + \tfrac{1}{2}\sum_{ij} A_A^+\delta_{ij} + \tfrac{1}{2}\sum_{ij} A_A^-(1-\delta_{ij}) \tag{44}$$

where B_A^l is the energy of electron i in an atomic spin-orbital in the field of the core with l the azimuthal quantum number, A_A^+ and A_A^- are the interaction energies between electrons with the same spin and opposite spin respectively, and δ_{ij} is the Kronecker delta.

If the electrons are in the same orbital this reduces to

$$E_{A,j} = n_{A,j}B_A^l + \tfrac{1}{2}n_{A,j}(n_{A,j}-1)A_A \tag{45}$$

where now $A_A^+ = A_A^- = A_A$ since the electron spins are necessarily opposite. If

$$b_A = B_A^l - \tfrac{1}{2}A_A \tag{46}$$

$$2c_A = A_A/2 \tag{47}$$

then Eq. (25) is obtained. This leads to the following expression for electronegativity

$$\chi_{A,j} = (B^1_A - \tfrac{1}{2} A_A) + (A_A/2)n_{A,j} \tag{48}$$

And in a like manner Iczkowski and Margrave's Eq. (21) can also be obtained. Thus, the fundamental constants in both schemes (and, therefore, also in Huheey's) can be related to theoretical quantities.

Implicit in Klopman's approach and recognized by both Iczkowski and Hinze is the fact that these relations represent non-continuous functions of $n_{A,j}$ and are, therefore, non-differentiable with respect to $n_{A,j}$. As pointed out by Ferreira $n_{A,j}$ is more reasonably interpreted in terms of a two electron bonding function[7]. In this case, the electron pair can be seen as spending $(n_{A,j}/2)\%$ of its time in the atomic orbital on A. In that sense the relations can be considered as differentiable. This viewpoint, of course, is in conformity with chemical intuition.

Klopman also showed that the electronegativity appears in a natural way in the diagonal matrix elements which result from the usual variational treatment of molecules[71]. In addition, he discussed the electronegativity equalization principle and noted that it should not only include atomic but also molecular terms to be in conformity with the variational principle[72]. He thus restated it in terms of the total molecular energy Em as

$$\frac{\partial Em}{\partial n^i_j} = \frac{\partial Em}{\partial n^i_k} \tag{49}$$

where n^i_j refers to the charge density due to electron i in atomic orbital j. Thus, the potential in each atomic orbital is the same throughout the molecule. Since the electronegativity depends on atomic and molecular terms, Klopman suggests the term "molecular orbital electronegativity". In addition, he refers to the orbitals as equipotential orbitals.

Recently, Klopman's work has been modified and extended to provide both a simplified calculational procedure for atomic or orbital electronegativity and also for group electronegativity[58]. It was found that a linear relation existed between Klopman's A_A and B^1_A values for each atom. This meant that (48) can be expressed in terms of A_A

$$\chi_{A,j} = A_A (0.5 + 0.17\, Ns_j) \tag{50}$$

where N is the number of valence electrons and s_j is the % s character in orbital j. It also meant that the ionization $(I_{A,j})$ could be expressed as

$$I_{A,j} = A_A (1 + 0.17\, Ns_j) \tag{51}$$

Noting that $I_{A,j}$ can also be expressed in terms of a screened nuclear charge and principal quantum number led to the following relation

$$\chi^0_{A,j} = 1.67\, G_j Z^2_{eff}/n^2_e + 0.41 \tag{52}$$

where G_j is a linear function of % p hybridization in the orbit j, n_e and Z_{eff} are Slater's screened effective principal quantum number and modified screened nuclear charge respectively. The constants put the scale into Pauling units. Values obtained from (52) were shown to correlate with other values very well over the entire periodic table including transition elements.

In addition, it was found that electronegativity can be written as

$$\chi_{A,j} = \chi^0_{A,j}\left(1 + 1.5 \sum_i \delta_{A,i}\right) \tag{53}$$

where the summation is over all bonded orbitals on A (including j). The 1.5 term is a result of the relations found between B^1_A and A_A and eliminates the need to estimate charge effects on electronegativity.

Equation (53) was used in conjunction with the bond electronegativity equalization principle of Hinze and the assumption of 2 electrons per bond to arrive at relations very similar to (38) and (39)

$$\sum_j \delta_{A,j} = \delta_A = (\chi/\chi_A\, 1.5) - 1 \tag{54}$$

$$\chi = (N + 1.5\, \delta_G)/\Sigma(\nu/\chi_A) \tag{55}$$

in which the symbols have the same meaning as above. Note that these relations are only true if each atom uses the same bonding orbital for each bond. Hence, the lack of reference to orbitals in the equations. It can be seen that for neutral groups (39) and (55) are equal. However, the atomic charges are different. As it stands, this method suffers from the same problems as all of the others mentioned thus far.

However, the fact that this method is based on an orbital point of view allows a way out of each of the difficulties. It is shown that by treating intra-atomic electron repulsion between valence orbitals in a more realistic fashion, Eq. (53) is modified to[73)]

$$\chi_{A,j} = \chi^0_{A,j}\left(1 + 0.5 \sum_{i \neq j} \delta_{A,i} + 1.5\, \delta_{A,j}\right) \tag{56}$$

This formula can be used with the bond electronegativity equalization and bond charge conservation principles to obtain values for the orbital charges. These latter two principles can be stated as follows. For each bond j between atoms A and B in the group or molecule

$$\chi_{A,j} = \chi_{B,j} \tag{57}$$

$$\delta_{A,j} + \delta_{B,j} = 0 \tag{58}$$

The use of relations (56), (57), and (58) leads to a resolution of the problems which plague all other simplified schemes. Values for group electronegativity obtained from this method have been compared to other schemes as well as empirical data with very good results. In addition, charge effects have been shown to be quite reasonable. Of course, Eq. (56) can also be used with other methods of calculating atomic electronegativity if desired, (i.e. not just Eq. (52)). In terms of the simplicity and ease of use criteria this scheme represents an advance over previous methods.

3.6 Other Work

Recently Ponec has introduced a generalization of the orbital electronegativity concept of Hinze, Whitehead, and Jaffe[52, 74]. The method is based on the semi-empirical CNDO approximation. Ponec's basic equation is

$$\chi_{A,J} = -U_{JJ}^{A} - (P_A - \tfrac{1}{2})\gamma_{AA} \tag{59}$$

where U_{jj}^{A} is the one electron energy of orbital j, γ_{AA} is an electron repulsion integral and P_A is the total electron density associated with atom A. For neutral atoms this reduces to Mulliken-Jaffe values. However, inia molecule a global electronegativity term (ξ_A) can be definad as

$$\xi_A = \Sigma p_{jj}\chi_{A,J}/\Sigma p_{JJ} \tag{60}$$

where p_{JJ} is the charge density in atomic orbital j on A. ξ_A also reduces to Mulliken-Jaffe electronegativity for isolated atoms. Values of ξ_A for some molecules were correlated to ESCA chemical shifts with good results. It was further shown that extension of these ideas to the INDO approximation was not possible without introducing further arbitrary assumptions[74]. It would be of interest to compare results obtained from one of the simpler methods discussed previously to results from (60).

Several workers[48, 59] have investigated the use of more general expressions for energy than the simple quadratic one discussed thus far. It was found that in some cases it may be necessary to include higher order terms[59].

Baird, Sichel, and Whitehead have suggested an approach to unifying the various electronegativity schemes which utilize an equalization principle[46]. Their work is based on simple MO theory. However, they include an assumption requiring that the total electronic energy of the molecule be partitioned as a sum of atomic energies. The usefulness of this work in the present context appears to be primarily in the possibility of providing a language to discuss the various schemes in a unified fashion if desired.

3.7 Group Electronegativity – Numerical Comparisons

Other methods have been developed in the past several years for the specific purpose of estimating electronegativities of organic groups[75, 76]. These are logically independent of the main trend in electronegativity theory development as represented by the work already reviewed. They can be viewed more in terms of "classical" work in electronegativity theory as applied to groups. The most important is due to Inamoto and collaborators[75, 76]. It is based on Gordy's method [Eq. (10)] but uses a different method to obtain Z'. The values from this scheme have been compared extensively to NMR data and other calculated and empirical values. Results indicate the method to be quite useful in correlating data.

A second method has been developed by Taft and co-workers[77, 78]. It utilizes a purely theoretical approach. The values are the atomic electron population on the hydrogen atom in HG compounds (where G is the group being considered). These are obtained from ab initio calculations using a Mulliken population analysis. This method has also been compared to NMR data with good results.

Table 2 presents a comparison of electronegativity values for ten of the important organic groups for which there is data available for the methods treated here. The "empirical" data has been taken from a review by Wells[4)]. The results represent what he considers to be the best available from empirical data. They have become something of a standard by which other scales are measured[57,73,79)]. The values in parentheses are less certain than the others.

Six scales are compared to the Wells data. Three represent complete electronegativity equalization schemes (Huheey, Bratsch, Sanderson), one a partial electronegativity equalization scheme (Mullay), and the two methods not using electronegativity equalization (Inamoto and Taft). Correlation coefficients were calculated relating each scale to the Wells data. The results indicate that total electronegativity equalization of the type described in the literature tends to overestimate the effects of atoms (or groups) bonded to the central atom of a group. Thus, some modification as, for example, in relation (56) is required in order to treat group electronegativity adequately. Since recent theoretical work indicates that electronegativity is equalized in a molecule or group[50,55)], it is apparent that greater clarification is needed in this area.

It is also readily apparent from the table that empirical data can be reasonably explained in terms of current treatments of electronegativity. This suggests that the phenomena being used to derive the data can be understood in terms of the simple concepts upon which the schemes are based. In the case of relation (56) these concepts include the current views of electronegativity as a potential and the use of electronegativity equalization in a bond.

3.8 Summary

As in the previous section, the work reviewed here does not represent all of the effort put into the development of the electronegativity concept during this period. However, as in

Table 2. Comparison of selected empirical and calculated group electronegativity values

	Empirical[a]	Sanderson[b]	Huheey[c]	Bratsch[b]	Mullay[d]	Inamoto[e]	Taft[f]
CH_3	2.3	2.33	2.27	2.23	2.32	2.14	0.17
COOH	(2.85)	2.86	3.52	2.80	3.15	2.36	0.18
CN	3.3	2.68	3.84	2.77	3.46	2.61	0.31
CF_3	3.35	3.47	3.46	3.49	3.10	2.47	0.25
NH_2	3.35	2.49	2.61	2.42	3.15	2.47	0.33
NO_2	3.4	3.29	4.83	3.30	4.08	2.75	0.40
OH	3.7	2.81	3.51	2.68	3.97	2.79	0.43
OCH_3	(3.7)	2.52	2.68	2.44	4.03	2.82	0.44
SiH_3	(2.2)	2.18	2.21	2.12	1.97	1.79	−0.13
SH	(2.8)	2.48	2.32	2.37	2.42	2.17	0.12
Correlation coefficient	–	0.54	0.54	0.52	0.92	0.94	0.91

[a] Ref. 4; [b] Ref. 57; [c] Ref. 56; [d] Ref. 73; [e] Ref. 75, 76; [f] Ref. 77

the case of atomic electronegativity scales, it reviews all of the previous work which is necessary for an appreciation of current efforts. Hopefully, it gives a good indication of one of the major directions along which work in this field has progressed. It can be seen that there are elements common to the various methods which can be used to point out the factors that are central to this area of electronegativity theory.

(1) Mulliken's definition of atomic electronegativity[10] is assumed to be better than the others developed in the earlier phase of electronegativity work. The only author that appeared to disagree is Sanderson. But even his work is not inconsistent with this point of view[37]. Thus, the fundamental authority for most of the recent work is Mulliken.

(2) The electronegativity of an atom is seen as being dependent on the environment of the atom. Specifically it depends on the charge induced in the atom by the other atoms bonded to it. Of course, along with this is the belief that the hybridization of the orbital used in bonding has a great effect on the electronegativity, as would also the oxidation state of the atom. It was recognized in previous work that the environment was important but the schemes developed were not able to easily account for the differences. Current methods can do this.

(3) It is generally taken for granted that an electronegativity equalization principle is in some sense valid. Work done by Parr and others suggest that the electronegativities of atoms and orbitals are totally equalized in a molecule or group[50, 55]. This is in line with Sanderson's original concept. At present, it appears that there are convincing arguments for both total electronegativity equalization as well as for partial equalization[36, 55, 73]. Since Parr's work is strictly valid for the ground state and natural orbitals it is not certain how this is related to other situations. The situation is sufficiently unclear that some authors continue to use a partial equalization scheme with good results[36, 73]. As shown above, recent work with group electronegativity indicates that there are definite disadvantages with models based on the concept of total equalization[4, 56, 73]. This seems to be an area requiring more work[36, 80].
The use of this idea has allowed the ionic character of the bonds and atomic charges to be calculated as a natural consequence of the electronegativity theory. In the past molecular charge distribution was seen as related to electronegativity but the two quantities were not calculable from one unified method[1, 81]. This is no longer true because of the use of electronegativity equalization.

(4) Conceptual simplicity and ease of use remain important in developing a reasonable electronegativity method. Once the basic concepts were developed in the earlier papers, later efforts were made to provide an easily used method. This is true for both atomic charge and group electronegativity.

(5) The techniques discussed in this section allow group electronegativity values to be calculated in a reasonable manner. Thus, recent investigations appear to place greater stress on this quantity.

4 General Overview

It can be seen that electronegativity theory has progressed significantly since Pauling's original work. The general trend appears to be toward a gradual appreciation of the various complexities involved in the concept. Thus, it has evolved from a one-number-per-element scale to a method involving a relation that yields a range of values for each atom, with the most recent relations accounting for at least valence state and partial charge on the atom. This, as pointed out by Batsanov[6)], is the usual state of affairs with scientific concepts, i.e. moving from a naive idea to a deeper understanding.

Throughout this growth, the desire to maintain simplicity has remained. And apparently along with this is the vision that the electronegativity concept is somehow best embodied in an easily used scheme. It is too early to predict how the recent efforts in "molecular electronegativity"[82, 83)] and the use of the chemical potential idea[62, 84, 85)] will modify these trends. From the present perspective, it appears that the marriage of the two viewpoints certainly shows promise in advancing chemical knowledge.

5 References

1. Pauling, L.: The Nature of the Chemical Bond, 3rd Ed., Cornell University Press, Ithaca, N.Y. 1960
2. Pritchard, H. O., Skinner, H. A.: Chem. Rev. *55,* 745 (1955)
3. Hinze, J.: Fortschritte der Chemischen Forschung: In: Theoretische Organische Chemie *9.3,* Springer, Berlin 1968
4. Wells, P. R.: In Progr. Phys. Org. Chem. (Streitweiser, A. Jr., Taft, R. W. Eds.), Wiley (Interscience), N.Y., 111, 1968
5. Huheey, J. E.: Inorganic Chemistry: Principles of Structure and Reactivity, 2nd Ed., Harper and Row 1978
6. Batsonov, S. S.: Russ. Chem. Rev. *37,* 332 (1968)
7. Ferreira, R.: Electronegativity and Chemical Bonding. In: Adv. Chem. Phys. *13* (Prigogine, I. Ed.), Wiley (Interscience), N.Y., 55, 1967
8. Van Vechten, J. A.: Proc. Int. Sch. Phys. "Enrico Fermi" *52,* 464 (1972)
9. Pauling, L.: J. Am. Chem. Soc. *54,* 3570 (1932)
10. Mulliken, R. S.: J. Chem. Phys. *2,* 782 (1934)
11. Sanderson, R. T.: J. Chem. Educ. *29,* 539 (1952); *31,* 2, 238 (1954)
12. Allred, A. L., Rochow, E. G.: J. Inorg. Nucl. Chem. *5,* 264 (1958)
13. Gordy, W.: Phys. Rev. *69,* 604 (1946)
14. Pauling, L.: J. Am. Chem. Soc. *53,* 1367, 3225 (1931); *54,* 988 (1932)
    Pauling, L., Yost, D. M.: Proc. Nat. Acad. Sci. *18,* 414 (1932)
15. Pauling, L., Sherman, J.: J. Am. Chem. Soc. *59,* 1450 (1937)
16. Allred, A. L.: J. Inorg. Nucl. Chem. *17,* 215 (1961)
17. Fung, B.: J. Phys. Chem. *69,* 596 (1965)
18. Haissinsky, M.: J. Phys. *7,* 7 (1946); J. Chem. Phys. *46,* 298 (1949)
19. Walsh, A. D.: Discuss. Faraday Soc. *2,* 18 (1947)
20. Daudel, P., Daudel, R.: J. Phys. *7,* 12 (1946)
21. Mulliken, R. S.: J. Chim. Phys. *46,* 497 (1949)
22. Moffit, W.: Proc. Roy. Soc. (London) *A202,* 548 (1950)
23. Sanderson, R. T.: Inorganic Chemistry, Reinhold Publishing Corp., N.Y. 1967
24. Hinze, J., Jaffe, H. H.: J. Am. Chem. Soc. *84,* 540 (1962); Can. J. Chem. *41,* 1315 (1963); J. Phys. Chem. *67,* 1501 (1963)
25. Hinze, J., Whitehead, M. A., Jaffe, H. H.: J. Am. Chem. Soc. *85,* 148 (1963)
26. Huheey, J. E.: J. Inorg. Nucl. Chem. *27,* 2127 (1965)
27. Boyd, R. J., Markus, G. E.: J. Chem. Phys. *75,* 5385 (1981)
28. Mande, G., Deshmukh, P., Deshmukh, P.: J. Phys. B *10,* 2293 (1977)
29. Zhang, Y.: Inorg. Chem. *21,* 3886 (1982)
30. Gordy, W.: J. Chem. Phys. *14,* 305 (1946)
31. Gordy, W.: ibid. *19,* 792 (1951)

32. Politzer, P., Parr, R. G., Murphy, D. R.: ibid. *79,* 3859 (1983)
33. Iczkowksi, R. P., Margrave, J. L.: J. Am. Chem. Soc. *83,* 3547 (1961)
34. Bratsch, S. G.: J. Chem. Educ. *61,* 588 (1985)
35. Ray, N. K., Samuels, L., Parr, R. G.: J. Chem. Phys. *70,* 3680 (1979)
36. Mortier, W. J., Van Gnechten, K., Gasteiger, J.: J. Am. Chem. Soc. *107,* 829 (1985)
37. Sanderson, R. T.: Chemical Bonds and Bond Energy, 2nd Ed., Academic Press, N.Y. 1976
38. Simons, G., Zandler, M. E., Talaty, E. R.: J. Am. Chem. Soc. *98,* 7869 (1976)
39. Frost, A. A.: J. Chem. Phys. *47,* 3707 (1967)
40. St. John, J., Bloch, A. N.: Phys. Rev. Lett. *33,* 1095 (1974)
41. Simons, G.: J. Chem. Phys. *55,* 756 (1971)
42. (a) Jørgensen, C. K.: Progress. Inorg. Chem. *12,* (1970) 101; (b) Jørgensen, C. K.: Structure and Bonding *24,* 1 (1975); *30,* 14 (1976)
43. Spiridonov, V. P., Tatevskii, V. M.: Russ. J. Phys. Chem. *37,* 522, 661, 848, 1070 (1963)
44. Sanderson, R. T.: Science *114,* 670 (1951)
45. Van Hooydunk, G.: Z. Naturforsch. *28a,* 933 (1973); *30a,* 223 (1975)
46. Baird, N. C., Sichel, J. M., Whitehead, M. A.: Theoret. Chim. Acta *11,* 38 (1968)
47. Ferriera, R., de Amorim, A. O.: ibid. *58,* 131 (1981)
48. Lohr, L. L., Jr.: Int. J. Quantum Chem. *25,* 211 (1984)
49. Bartolotti, L. J., Gadre, S. R., Parr, R. G.: J. Am. Chem. Soc. *102,* 2945 (1980)
50. Politzer, P., Weinstein, H.: J. Chem. Phys. *71,* 4218 (1979)
51. Voigt, B., Dahl, J. P.: Acta Chem. Scand. *A26,* 2923 (1972); *A28,* 1043, 1068 (1974)
52. Ponec, R.: Coll. Czech. Chem. Comm. *47,* 1479 (1982)
53. Jolly, W. L., Perry, W. B.: J. Am. Chem. Soc. *95,* 5442 (1973); Inorg. Chem. *13,* 2686 (1974)
54. Ohwada, K.: Polyhedron *2,* 853 (1984)
55. Parr, R. G., Donnelly, R. A., Levy, M., Palke, W. E.: J. Chem. Phys. *68,* 3801 (1978)
56. Huheey, J. E.: J. Phys. Chem. *69,* 3284 (1965); *70,* 2086 (1966)
57. Bratsch, S. G.: J. Chem. Educ. *62,* 101 (1985)
58. Mullay, J.: J. Am. Chem. Soc. *106,* 5842 (1984)
59. Watson, R. E., Bennett, L. H., Davenport, J. W.: Phys. Rev. B *27,* 6428 (1983)
60. Pritchard, H. O.: J. Am. Chem. Soc. *85,* 1876 (1963)
61. Evans, R. S., Huheey, J. E.: J. Inorg. Nucl. Chem. *32,* 373, 383, 777 (1970)
62. Reed, J. L.: J. Phys. Chem. *85,* 148 (1981)
63. Whitehead, M. A., Baird, N. C., Kaplansky, M.: Theoret. Chim. Acta *3,* 135 (1965)
64. Sichel, J. M., Whitehead, M. A.: ibid. *5,* 35 (1966)
65. Huheey, J. E.: J. Org. Chem. *36,* 204 (1971)
66. Parr, R. G., Pearson, R. G.: J. Am. Chem. Soc. *105,* 7512 (1983)
67. Huheey, J. E.: J. Org. Chem. *31,* 2365 (1966)
68. Evans, R. S., Huheey, J. E.: Chem. Phys. Lett. *19,* 114 (1973)
69. Hancock, J. R., Hardstaff, W. R., Johns, P. A., Langler, R. F., Mantle, W. S.: Can. J. Chem. *61,* 1472 (1983)
70. Jardine, W. K., Langler, R. F., MacGregor, J. A.: ibid. *60,* 2069 (1982)
71. Klopman, G.: J. Am. Chem. Soc. *86,* 1463, 4550 (1964); *87,* 3300 (1965)
72. Klopman, G.: J. Chem. Phys. *43,* 5124 (1965)
73. Mullay, J.: J. Am. Chem. Soc. *107,* 7271 (1985); *108,* 1770 (1986)
74. Ponec, R.: Theoret. Chim. Acta *59,* 629 (1980)
75. Inamoto, N., Masuda, S.: Tetrahedron Lett. 3287 (1977)
Inamoto, N., Masuda, S., Tori, K., Yoshimura, Y.: ibid. 4547 (1978)
76. Inamoto, N., Masuda, S.: Chem. Lett. *1003,* 1007 (1982)
77. Reynolds, W. F., Taft, R. W., Topsom, R. D.: Tetrahedron Lett. 1055 (1982)
78. Marriott, S., Reynolds, W. F., Taft, R. W., Topsom, R. D.: J. Org. Chem. *49,* 959 (1984)
79. Reynolds, W. F.: J. Chem. Soc., Perkin Trans. *2,* 985 (1980)
80. Gasteiger, J., Marsili, M.: Tetrahedron Lett. 3181 (1978); Tetrahedron *36,* 3219 (1980)
81. Barbe, J.: J. Chem. Educ. *60,* 640 (1983)
82. Bohm, M. C., Sen, K. D., Schmidt, P. C.: Chem. Phys. Lett. *78,* 357 (1981)
83. Vera, L., Zuloaga, F.: J. Phys. Chem. *88,* 6415 (1984)
84. Berthier, G.: Adv. Quantum Chem. *8,* 183 (1974)
85. Jørgensen, C. K.: Orbitals in Atoms and Molecules, Academic Press, N.Y. 1962

Absolute Electronegativities as Determined from Kohn-Sham Theory

Libero J. Bartolotti

Department of Chemistry, University of Miami, Coral Gables, Florida 33124, U.S.A.

Recent developments in the density functional formalism of absolute electronegativities are presented. The electronegativity has been obtained from the Kohn-Sham Lagrange multipliers via a transition state method and a non-transition state method. The calculations discussed here have employed both non-spin-polarized and spin-polarized approximate Kohn-Sham theory to obtain the Lagrange multipliers.

I. Introduction 28

II. The Density Functional Formalisms 29

III. Electronegativity Formula 31

IV. Numerical Results 33
 A. Atoms 33
 B. Molecules 38

V. Conclusion 39

VI. References 39

Structure and Bonding 66

I. Introduction

Electronegativity, in one form or another, has long been one of the most useful concepts in chemistry. Until recently, however, a quantitative description of electronegativity has been best understood outside the framework of quantum mechanics. The connection between electronegativity χ and quantum mechanics was made when Parr, Donnelly, Levy and Palke[1] identified it as the negative of the chemical potential μ of density functional theory[2-6]. In density functional theory, the chemical potential is introduced as the Lagrange multiplier which insures that the number of particles will be conserved in the minimization of the energy. It is equal to the functional derivative with respect to the density of the energy density functional.

$$\mu = \left(\frac{\delta E[\varrho]}{\delta \varrho}\right)_V \tag{1}$$

It can also be written as the derivative of the ground state energy with respect to the number of electrons,

$$\mu = \left(\frac{\partial E(N)}{\partial N}\right)_V = -\chi \tag{2}$$

Here V is the nuclear potential and it is held constant. The electronegativity (as defined by Eqs. (1) and (2)) is a characteristic property of an atom or molecule and if two or more species of different electronegativities come together to form a new species, a common electronegativity results (Sanderson's equalization principle[7])[1, 8]. It is also important to note that this definition of the electronegativity is consistent with Mulliken's definition[9, 10].

$$\chi = \frac{I + A}{2} \tag{3}$$

where I is the ionization potential and A is the electron affinity.

Certain aspects of Eq. (2) are not new. Some current electronegativity theories have treated the energy of an atom or orbital as a differentiable quadratic function of the number of electrons[11-20] and have defined the electronegativity as the negative of the slope of this energy as a function of N. Also, Gyftopoulos and Hatsopoulos[21] had previously equated the electronegativity to the negative of the chemical potential in their statistical ensemble theory of electronegativity.

Since the Hohenberg-Kohn energy density functional theory[2] and its generalization[3-6] were proven for closed systems with a fixed, integer number of electrons, Eq. (2) raises two important questions: (i) Is the energy defined for a non-integral number of electrons? (ii) Is the energy differentiable with respect to the number of electrons? Perdew and coworkers[22, 23] have shown that the answers to both questions are yes. See also the paper by Parr and Bartolotti[24].

The above provides the motivation and a formal justification for performing actual calculations to determine absolute electronegativities of chemical systems. We will describe below what has been done in this area with an emphasis on our calculations of atomic electronegativities within the density functional formalism.

II. The Density Functional Formalisms

The ground state electronic energy as a function of density can be written as follows

$$E[\varrho] = T[\varrho] + J[\varrho] + K[\varrho] + V_{ne}[\varrho], \tag{4}$$

where

$$J[\varrho] = \frac{1}{2} \int\int \frac{\varrho(1)\,\varrho(2)}{r_{12}}\, d\tau_1\, d\tau_2 \tag{5}$$

is the Coulomb energy density function,

$$V_{ne}[\varrho] = \int \varrho v d\tau \tag{6}$$

is the nuclear-electron energy density functional, $T[\varrho]$ is the kinetic energy density functional and $K[\varrho]$ is the exchange-correlation energy density functional. These latter two functionals are as yet unknown. The density is found by minimizing $E[\varrho]$ with respect to τ for a given ϱ subject to the constraint

$$N = \int \varrho d\tau \tag{7}$$

i.e.

$$\delta\,\{E[\varrho] - \mu\,(\int \varrho d\tau - N)\} \tag{8}$$

where μ (the chemical potential) is a Lagrange multiplier. This variation leads to the Euler-Lagrange equation [Eq. (1)].

Although Eq. (1) formally provides a means to obtain ϱ and μ, it is not used in practice. The unknown $E[\varrho]$ presents the major problem. In 1965, Kohn and Sham[25] showed that one could obtain μ without a knowledge of the whole of $E[\varrho]$. They partitioned the ground state electronic energy as follows:

$$E[\varrho] = T_s[\varrho] + J[\varrho] + E_{xc}[\varrho] + V_{ne}[\varrho] \tag{9}$$

where

$$E_{xc}[\varrho] = K[\varrho] + T[\varrho] - T_s[\varrho] \tag{10}$$

They then described an orbital prescription for obtaining $T_s[\varrho]$, which is defined as[3, 25–27]

$$\begin{aligned} T_s[\varrho] &= \min\left(\frac{1}{8}\sum_{i=1}^{N} \int |\vec{\nabla}\varrho_i|^2/\varrho_i d\tau\right) \\ &= \min\left\{-\frac{1}{2}\sum_{i=1}^{N} n_i \int \phi_i^* \nabla^2 \phi_i d\tau\right\} \end{aligned} \tag{11}$$

where

$$\varrho = \sum_{i=1}^{N} \varrho_i = \sum_{i=1}^{N} n_i |\phi_i|^2 \tag{12}$$

and

$$\int \varrho_i d\tau = n_i \langle \phi_i | \phi_i \rangle = n_i = 1 \tag{13}$$

The $\{\phi_i\}$ are the Kohn-Sham orbitals, the ϱ_i are the Kohn-Sham orbital densities and the (n_i) are the occupation numbers of the $\{\phi_i\}$. The functional $T_s[\varrho]$ is obtained by searching over the $\{\phi_i\}$ which satisfies Eqs. (12) and (13) and then evaluating the integral in Eq. (11). The minimum value of this integral is assigned to $T_s[\varrho]$. This minimization procedure is equivalent to solving the following differential equation:

$$-\frac{1}{2} \frac{\nabla^2 \varrho_i^{1/2}}{\varrho_i^{1/2}} + \frac{\delta E_{xc}[\varrho]}{\delta \varrho} + v^* = \varepsilon_i \tag{14}$$

Here v^* is the classical electrostatic potential and it is defined as

$$v^* = \frac{\delta J[\varrho]}{\delta \varrho} + v \tag{15}$$

The Lagrange multipliers ε_i are related to the energy by

$$(\partial E / \partial n_i)_{(n_j)_{i \neq j}} = \varepsilon_i \tag{16}$$

When taking this derivative all the n_j are kept constant except for n_i. To simplify notation, we will no longer explicitly show that these quantities are held constant in taking the derivative.

Equation (16) was derived in Xα theory by Slater[28]. More recently, Janak[29] has shown that $\frac{\partial E}{\partial n_i} = \varepsilon_i$ is valid for any $E_{xc}[\varrho]$ and that it includes orbital relaxation. This equation is the key formula for our application.

Slater[30, 31] has equated the $\partial E/\partial n_i = \varepsilon_i$ in Xα theory with the orbital electronegativites as defined by Hinze, Whitehead and Jaffe[13]. However, the electronegativity as defined in the next section is not an orbital electronegativity, but the electronegativity of the atom or molecule.

The method of Kohn-Sham is in principle exact. Unfortunately, the exact form of the exchange-correlation functional $E_{xc}[\varrho]$ is not exactly known and approximations to it must be used in solving for $\{\phi_i\}$ in Eq. (14). Reasonable approximations to it are known and have been successfully used in atomic and molecular calculations[27, 28, 32]. Some recent review articles discuss energy density functional and/or Kohn-Sham theory in great detail[32–36].

III. Electronegativity Formula

We will now use the results of the last section to derive an expression for the electronegativity in terms of the Kohn-Sham Lagrange multipliers. The derivation presented here was used in our recent spin-polarized calculations of absolute electronegativities[37)] and differs from the transition state concept used by us earlier[38)]. In order to keep the derivation and resulting equations as simple as possible, we will limit our derivation to the non-spin-polarized case. The extension to the spin-polarized case is straight forward.

The arguments used in the present derivation are similar to those used by Slater and Wood[28, 39)] in the derivation of the transition state method for calculating ionization potentials within Xα theory. First, a Taylor series expansion of the energy is taken about a reference state "O", i.e.

$$E = E_0 + \sum_{i} (n_i - n_{i0}) \left.\frac{\partial E}{\partial n_i}\right|_0 + \frac{1}{2!} \sum_{i,j} (n_i - n_{i0})(n_j - n_{j0}) \left.\frac{\partial^2 E}{\partial n_i \partial n_j}\right|_0 + \frac{1}{3} \sum_{i,j,k} (n_i - n_{i0})(n_j - n_{j0})(n_k - n_{k0}) \left.\frac{\partial^3 E}{\partial n_i \partial n_j \partial n_k}\right|_0 + \ldots\ldots \quad (17)$$

where $|_0$ denotes that the derivatives are evaluated with respect to the reference state. Since we are interested in the χ of neutral atoms and molecules, the reference state is the neutral system. Secondly, we consider what happens to the $\{n_i\}$ when the atom or molecule goes from the positive ion to the negative ion. We construct $E(N - \delta)$ from Eq. (17) by removing a net of δ electrons from the N_i involved in ionization. Likewise, we construct $E(N + \delta)$ by adding a net of δ electrons to the n_i which are involved in going from the neutral atom to the negative ion. If only a net of two orbitals are involved (the maximum number generally encountered), we find

$$E(N + \delta) - E(N - \delta) = \delta \left(\left.\frac{\partial E}{\partial n_i}\right|_0 + \left.\frac{\partial E}{\partial n_j}\right|_0 \right) + O(\delta^3) \quad (18)$$

The expression for the electronegativity can now be found by taking the limit

$$\chi = -\left(\frac{\partial E}{\partial N}\right)_V = -\lim_{\delta \to 0} \frac{E(N + \delta) - E(N - \delta)}{2\delta}$$
$$= -\frac{1}{2} \left(\left.\frac{\partial E}{\partial n_i}\right|_0 + \left.\frac{\partial E}{\partial n_j}\right|_0 \right) = \frac{1}{2} (\varepsilon_i + \varepsilon_j) \quad (19)$$

In other words, $(\partial E/\partial N)_V$ is taken as the average of the left and right derivatives of $E(N)$. This is consistent with Mulliken's formula. In obtaining Eq. (19) we need only assume that $E(N)$ is a smooth function about $N + \delta$ electrons and about $N - \delta$ electrons. Note that ε_i and ε_j are obtained from a ground state calculation and that when orbital ϕ_i is degenerate with orbital ϕ_j, the electronegativity is given by $\chi = -\varepsilon_i$.

If we do not take the limit in Eq. (19) and let δ equal a half, we recover our earlier transition state method for calculating the electronegativity[38)]. When two different orbitals are involved in going from the positive to the negative ions in the transition state method the reference state is one where a half an electron is in the ith orbital and a half an electron is in the jth orbital. If the two orbitals are degenerate, the transition state is generally (but not always) the ground state. The transition state was introduced to artificially make E(N) a smooth function of N for closed shell and closed subshell neutral systems. In the transition state method, third- and higher-order derivatives [terms of order δ^3 in Eq. (18)] appear in the expression for χ [unlike Eq. (19)]. It was found, however, that these terms are small and can be neglected[38]. This is also true for transition state calculations of ionization potentials and electron affinities[28]. The transition state equivalent of Eq. (19) is

$$\chi \approx -\frac{1}{2}(\varepsilon_i^t + \varepsilon_j^t) \tag{20}$$

where ε_i^t and ε_j^t are transition state Lagrange multipliers. The ionization energy of the ith orbital is given by Slater's transition state method as

$$I \approx -\left.\frac{\partial E}{\partial n_i}\right|_{n_i = 1/2} = -\varepsilon_i^t \tag{21}$$

where a half an electron has been removed from the ith orbital and the calculation is performed with a 1/2 charge on the chemical species. Likewise, the electron affinity of the jth orbital is given by

$$A \approx -\left.\frac{\partial E}{\partial n_j}\right|_{n_j = 1/2} = -\varepsilon_j^t \tag{22}$$

where a half an electron has been added to the jth orbital and the calculation is performed with a $-1/2$ charge on the chemical species. Equations (20)–(22) are written as approximations since the third- and higher-order derivatives are neglected. Mulliken's definition for the electronegativity is also implied by Eqs. (20)–(22).

Unfortunately, the widely employed approximations to $E_{xc}[\varrho]$ rarely lead to numerical solutions for negative ions (even slightly negative ions) and we cannot use Eq. (22) to estimate the electron affinity for these systems. These approximate $E_{xc}[\varrho]$ fail because their functional derivatives, $\delta E_{xc}[\varrho]/\delta\varrho$, behave incorrectly far from the nuclei[27, 40]. An exception is the self-interaction correction approximation to $E_{xc}[\varrho]$[27, 40]. The electronegativity can, however, be determined from Mulliken's formula, namely

$$A = 2\chi - I \tag{23}$$

The concept of hardness introduced by Pearson[41] has recently been quantized by Parr and Pearson[42]. They define the hardness as

$$\eta = \frac{1}{2}(\partial^2 E/\partial N^2)_V \tag{24}$$

which is given in terms of I and A by

$$\eta = \frac{1}{2}(I - A) \tag{25}$$

Therefore, the absolute hardness of a chemical system can be calculated using Eq. (25) and the ionization potentials and electron affinities determined above or it can be determined directly from Eq. (24).

The transition state concept requires that half an electron be placed in certain orbitals and raises an important question. We know how to write $E[n_i, \phi_i]$ (actually only approximations to it) for integer n_i. For non-integer n_i we usually use the same expression for E. Is this the correct way to proceed? Gopinathan and Whitehead[43)] have presented some covincing arguments that within Hyper-Hartree-Fock theory and Xα theory it is not the best way to proceed. They found that one should use a modified expression for E and that E(N) as a function of N is a series of broken straight lines. However, as also shown by them, the results obtained by both approaches lead to the same results. Perdew and coworkers[22, 23)] have recently shown that the E(N) vs. N curve for the exact E is also a series of broken straight lines.

IV. Numerical Results

A. Atoms

The equations of the last section have been used to calculate absolute electronegativities[37, 38, 44–48)], electron affinities[37, 38, 44, 48, 49)], and absolute hardnesses[37, 48)] of atomic systems. The first such calculations were reported by Bartolotti, Gadre and

Table 1. Absolute electronegativities in eV for H through Ar

Atom	a	b	c	d
H	7.97		5.27	5.74
He	12.61	12.93	7.93	8.00
Li	2.58	2.58	1.69	2.74
Be	3.80	4.16	3.52	4.03
B	3.40	4.08	4.08	4.37
C	5.13	6.44	6.39	6.52
N	6.97	7.97	5.78	6.67
O	8.92	9.61	6.45	7.67
F	11.00	11.13	9.85	10.76
Ne	10.31	10.47	6.60	6.96
Na	2.32	2.32	1.67	2.73
Mg	3.04	3.27	2.56	3.22
Al	2.25	2.69	2.70	3.24
Si	3.60	4.40	4.39	4.91
P	5.01	3.53	4.38	5.41
S	6.52	6.88	5.18	6.39
Cl	8.11	8.17	7.50	8.53
Ar	7.11	7.25	4.93	5.49

a Taken from Ref. 38. Non-spin-polarized transition state Xα calculation
b Taken from Ref. 47. Spin-polarized transition state Xα calculation
c Taken from Ref. 37. Spin-polarized non-transition state Xα calculation
d Taken from Ref. 37. Spin-polarized non-transition state calculation. The Gunnarsson-Lundqvist approximation to $E_{xc}[\varrho]$ was used

Parr[38]. They used the non-spin-polarized transition state method to calculate the electronegativities and electron affinities for the atoms hydrogen through xenon. The Xα approximation to $E_{xc}[\varrho]$ was used in Eq. (14). The α parameters used were those derived by Schwartz[50,51] for ground state atoms such that the virial ratio was satisfied when the Xα orbitals were inserted into the Hartree-Fock expressions for the energy components. Their calculated χ are given in column two Tables 1 and 2 and in columns two and six of

Table 2. Absolute electronegativities in eV for K through Kr

Atom	[a]	[b]	[c]	[d]
K	1.92	1.92	1.47	2.38
Ca	1.86	2.05	2.48	3.24
Sc	2.52	2.90	3.40	4.17
Ti	3.05	3.01	4.16	4.92
V	3.33	3.11	4.09	[e]
Cr	3.45	3.21	2.30	3.69
Mn	4.33	3.56	3.38	4.70
Fe	4.71	3.67	4.41	5.62
Co	3.76	3.77	4.84	5.74
Ni	3.86	3.86	5.00	5.93
Cu	3.95	3.95	3.76	[e]
Zn	3.66	3.89	3.00	3.84
Ga	2.11	2.54	2.54	3.21
Ge	3.37	4.10	4.10	4.79
As	4.63	5.06	4.08	5.24
Se	5.91	6.22	4.79	6.09
Br	7.24	7.29	6.74	7.92
Kr	6.18	6.31	4.36	5.05

[a] Taken from Ref. 38. Non-spin-polarized transition state calculation
[b] Taken from Ref. 47. Spin-polarized transition state Xα calculation
[c] Taken from Ref. 37. Spin-polarized non-transition state calculation
[d] Taken from Ref. 37. Spin-polarized non-transition state calculation. The Gunnarsson-Lundqvist approximation to $E_{xc}[\varrho]$ was used
[e] Numerical convergence not reached for this atom

Table 3. Absolute electronegativities in eV for Rb through Xe

Atom	[a]	[b]	[c]	Atom	[a]	[b]	[c]
Rb	1.79	1.41	2.30	Pd	3.52	2.40	3.24
Sr	1.75	1.98	2.80	Ag	3.55	3.39	4.38
Y	2.25	2.59	3.41	Cd	3.35	2.80	3.66
Zr	3.01	3.63	4.50	In	2.09	2.48	3.18
Nb	3.26	2.30	3.64	Sn	3.20	3.85	4.57
Mo	3.34	2.30	3.69	Sb	‹.?7	3.84	4.98
Tc	4.58	3.72	5.05	Te	5.35	4.43	5.71
Ru	3.45	3.11	4.11	I	6.45	6.04	7.22
Rh	3.49	3.23	4.22	Xe	5.36	3.85	4.56

[a] Taken from Ref. 38. Non-spin-polarized transition state Xα calculation
[b] Taken from Ref. 37. Spin-polarized non-transition state Xα calculation
[c] Taken from Ref. 37. Spin-polarized non-transition state calculation. The Gunnarsson-Lundqvist $E_{xc}[\varrho]$ was used

Table 4. Absolute electronegativities in eV for Cs through Rn

Atom	a	b	c	Atom	a	b	c
Cs		1.29	2.13	Lu	3.87	2.34	3.27
Ba		2.30	3.14	Hf		3.71	4.60
La		2.93	3.72	Ta		4.15	5.13
Ce	1.80	2.97	3.77	W		4.81	5.79
Pr	1.84	3.52	4.34	Re		3.60	5.00
Nd	1.90	4.00	4.80	Os		4.62	5.93
Pm	1.93	4.41	5.20	Ir		5.55	6.79
Sm	1.94	4.79	5.56	Pt		3.44	4.47
Eu	1.95	2.33	3.19	Au		3.51	4.53
Gd	2.02	5.67	6.94	Hg		2.80	3.70
Tb	2.00	3.29	4.42	Tl		2.42	3.16
Dy	2.01	3.70	4.80	Pb		3.70	4.47
Ho	2.03	4.08	5.14	Bi		3.69	4.86
Er	2.04	4.42	5.45	Po		4.23	5.52
Tm	2.06	4.73	5.73	At		5.69	6.91
Yb	1.85	1.93	2.83	Rn		3.60	4.35

[a] Taken from Ref. 44. Relativistic transition state Xα calculation

[b] Taken from Ref. 37. Spin-polarized non-transition state Xα calculation

[c] Taken from Ref. 37. Spin-polarized non-transition state calculation. The Gunnarsson-Lundqvist $E_{xc}[\varrho]$ was used

Table 3. Most of the calculated χ involved only one Lagrange multiplier, $\chi = -\varepsilon_i^t$. For most of these cases, the transition state was the ground state. The exceptions to this were associated with the transition metal atoms. The transition state is dictated by how one goes from the positive ion to the negative ion and not the ground state electronic configuration. Thus, the electronegativity of vanadium was calculated with the electronic configuration [Ar] $4s^1 3d^4$; not the ground state configuration [Ar] $4s^2 3d^3$. The electronic configurations of the positive ion and negative ion were taken from spectroscopic data[52–54]. We will not reproduce their calculated electron affinities here.

Using relativistic Xα theory, Sen, Schmidt and Weiss[44] calculated the electronegativities and electron affinities of the rare earth atoms cerium through lutetium. They used the α parameters derived by Schwartz[50, 51] and also used the transition method to calculate the χ. Their calculated values of χ are listed in columns two and six of Table 4.

Spin-polarized transition state Xα calculations of electronegativities of the atoms helium through krypton have been reported by Manoli and Whitehead[47]. Their results are given in column three of Tables 1 and 2. That makes their work most interesting is their derivation of Eq. (20). By expanding the $\varrho^{1/3}$ term in the Xα potential for the positive ion density ϱ_+ and for the negative ion density ϱ_- and then neglecting the terms which were small, they found

$$\varrho_+ = \varrho_0^{1/3} \qquad (26)$$

where ϱ_0 is the density of the neutral atom. This was then used to derive

$$\varepsilon_i^t = -I + \frac{1}{2}\langle ji||ji\rangle \qquad (27)$$

and

$$\varepsilon_j^t = -A - \frac{1}{2}\langle ji||ji\rangle \tag{28}$$

where $\langle ji||ji\rangle$ is a coulomb integral for orbitals ϕ_i and ϕ_j. The ε_i^t and ε_j^t are transition state Lagrange multipliers determined when $n_i = 1/2$ and $n_j = 1/2$. Taking one half of the sum of Eqs. (27) and (28) and equating the result to Mulliken's formula for χ they derived Eq. (20). Neglecting the terms which were small in the derivation of Eq. (26) is apparently equivalent (numerically) to neglecting the third- and higher-order derivatives in the transition state derivation of Eq. (20). Their work reinforces the connection between Eq. (20) and Mulliken's definition of electronegativity.

Spin-polarized Xα and hyper-Hartree-Fock calculations of electronegativities, hardness, ionization potentials and electron affinities for the open-shell atoms lithium through fluorine, chlorine, bromine, and iodine have been performed by Gázquez and Ortiz[48]. For these open shell systems the transition state methods, (Eq. 20), are equivalent. Their Xα electronegativities were the same as the ones obtained by Bartolotti, Gadre and Parr[38]. They calculated the hardness directly from Eq. (24) and then used their calculated χ and η to determine the ionization potential

$$I = \chi + \eta \tag{29}$$

and the electron affinity

$$A = \chi - \eta \tag{30}$$

The advantage of their approach is that all four quantities (χ, η, I and A) can be obtained from a single ground state calculation. The disadvantage is that the hardness obtained is an unrelaxed hardness and that $(\partial^2E/\partial N^2)$ is easily calculated only for open shell systems. These latter problems were overcome when they showed that the hardness could be estimated by

$$\eta = \frac{1}{4}\langle r^{-1}\rangle \tag{31}$$

The calculated ionization potentials and electron affinities were very close to those predicted by Eqs. (21) and (23). Another important result of the paper by Gazquez and Ortiz[48] was the hyper-Hartree-Fock electronegativities. The important quantity in the expression for χ is $\partial E/\partial n_i$. This derivative is given in the hyper-Hartree-Fock theory for open shell systems by

$$\frac{\partial E}{\partial n_i} = \varepsilon_i + \frac{1}{2}\langle ii||ii\rangle \tag{32}$$

which can be used to write

$$\chi = -\varepsilon_i - \frac{1}{2}\langle ii||ii\rangle \tag{33}$$

Equation (32) has a long history[28, 43, 55–58]. Electronegativities calculated from Eq. (33) by Gazquez and Ortiz[48] are listed in Table 5 where they are compared to the corresponding $X\alpha$ values. The hyper-Hartree-Fock values of χ are all smaller than the corresponding $X\alpha$ values.

Finally, we discuss the spin-polarized non-transition state calculations of electronegativities, electron affinities and hardnesses recently reported by Robles and Bartolotti[37]. Their calculated χ for the atoms hydrogen through radon are given in the last two columns of Tables 1 and 2 and columns three, four, seven and eight of Tables 3 and 4. The $X\alpha$ approximation to $E_{xc}[\varrho]$ and the Gunnarsson-Lundqvist[59] approximation to $E_{xc}[\varrho]$ were used in their calculations. All calculations of χ were performed using the spectroscopic ground state configurations[52–54] and the Lagrange multipliers used in Eq. (19) were determined from the spectroscopic configurations of the positive ion and negative ion. In every case, there were at least two Lagrange multipliers involved in the calculations and only a few calculations involved one orbital. When two spin orbitals were involved, they consisted of one occupied orbital and one unoccupied orbital (regular case), two occupied orbitals (irregular case) and two unoccupied orbitals (most irregular case.)

Now that we have briefly discussed the various calculations of atomic electronegativities, we will make some general comparisons. From Tables 1 and 2 we see that spin-polarization raises the values of χ except for the transition metal atoms. Spin-polarized calculated ionization potentials are in general larger than their non-spin-polarized counterparts. Thus the spin-polarized electron affinities for the transition metals must be much smaller than the corresponding non-spin-polarized ones. In making this comparison we note that Manoli and Whitehead[47] used different values of α than did Bartolotti, Gadre and Parr[38]. But as pointed out by Manoli and Whitehead[47], the different values of α had very little effect on the results. Comparing columns three and four of Table 1 and 2 we see the effect of using then non-transition state method, Eq. (19), as opposed to the transition state method, Eq. (20). Electronegativities determined with Eq. (19) are generally smaller, again the transition metal atoms are the exception. This overall lowering of χ is welcome, especially for the closed shell systems. The erratic behavior in the transition metals is most likely due to the different electronic configurations used in the calculations. The transition state for the transition metals is

Table 5. Absolute electronegativities in eV using hyper-Hartree-Fock theory

Atom	a	b
Li	2.15	2.58
B	2.95	3.41
C	4.11	5.14
N	5.43	6.98
O	6.91	8.94
F	8.59	11.01
Cl	7.32	8.12
Br	6.83	7.25
I	6.26	6.46

[a] Taken from Ref. 48. Hyper-Hartree-Fock calculation
[b] Taken from Ref. 38. Non-spin-polarized transition state calculation

not, in general, the ground state. However, the non-transition state calculations involve the ground state electronic configuration. The results in Tables 1–4 also show that the Gunnarsson-Lundqvist[59)] $E_{xc}[\varrho]$ leads to large values of χ as well as larger values of I and A. Table 4 shows that the relativistic Xα transition state values of χ are smaller than the spin-polarized non-transition state χ. There are too many differences in the calculations to make a good comparison. However, we note that the electron affinities determined by Sen, Schmidt and Weiss[44)] were all negative. The electron affinities determined by Robles and Bartolotti[37] for these same atoms are positive, except for ytterbium and lutetium. This is in agreement with the self-interaction-correction finite-difference calculations of Cole and Perdew[40].

B. Molecules

There have not been any direct calculations of molecular electronegativities using either Eq. (19) or (20). However, one could use electronegativity equalization arguments to obtain an expression for the molecular electronegativity in terms of their atomic parts. This has been done by Gázquez and Ortiz[48)]. They derived the weighted arithmetic-mean formula

$$\chi^{AB} = \frac{\chi_A^0 \eta_B^0 + \chi_B^0 \eta_A^0}{\eta_A^0 + \eta_B^0} = -\frac{\varepsilon_A^0 \langle r^{-1} \rangle_B + \varepsilon_B^0 \langle r^{-1} \rangle_A}{\langle r^{-1} \rangle_A + \langle r^{-1} \rangle_B} \tag{34}$$

Table 6. Absolute electronegativities in eV for several AB molecular systems

Atom	a	b	c	d
HF	8.79	11.02	8.62	9.30
HCl	7.88	8.54	7.69	8.06
HBr	7.49	8.01	7.37	7.53
HI	6.97	7.27	6.95	6.98
BrF	8.74	9.19	8.89	8.49
ClF	9.22	9.65	9.29	9.18
ICl	7.51	7.43	7.49	7.14
BrCl	8.00	6.91	7.94	7.64
LiH	4.17	4.17	4.63	3.99
NaH	4.07	3.72	4.50	3.69
KH	3.57	3.30	4.16	3.15
SO	6.78	8.59	8.45	7.42
SeO	7.18	8.26	8.22	6.94
TeO	6.27	7.85	7.94	6.44
PN	6.55	7.51	7.59	5.74
AsN	6.31	7.35	7.50	5.46
SbN	5.72	6.62	6.94	5.14

[a] Taken from Ref. 60. Charge transfer-electronegativity equalization, using experimental first ionization potential and electron affinity

[b] Taken from Ref. 60. Based on a simple bond charge model, using experimental equilibrium bond lengths

[c] Taken from Ref. 60. Based on the geometric mean approximation, using Mulliken's electronegativity determined from the experimental first ionization potential and electron affinity

[d] Taken from Ref. 48. Calculated from atomic Xα transition state and Eq. (34)

and used it to estimate molecular electronegativities using their Xα calculated values of χ and $\eta = \langle r^{-1} \rangle /4$. Here, χ_A and η_A^0 represent respectively the electronegativity and hardness of the neutral atom A in its ground state. We reproduce their Table VI in our Table 6. The electronegativities predicted by Eq. (34) are quire reasonable.

V. Conclusion

The results of the last section show that one can use the density functional formalism to obtain reasonable values for absolute electronegativities. The calculated atomic χ all exhibit very similar trends and they are in agreement with the trends of empirical electronegativities[61]. The non-transition state calculations gave the best values of χ. So far, only two different approximations to $E_{xc}[\varrho]$ have been used in the electronegativity calculations. There are various ways to parameterize the local spin density approximation[32]. The Gunnarsson-Lundqvist parameterization[59] is just one way. Ceperley and Alder[62] have recently provided a very accurate parameterization. The self-interaction correction approximation[27] to $E_{xc}[\varrho]$ provides a very interesting alternative as does the atomic gradient expansion[63] of $E_{xc}[\varrho]$. Which approximation yields the best results? What is the relativistic contribution to χ for the heavy atoms?

Calculated absolute electronegativities of molecules and molecular fragments are completely lacking. Such calculations are very important and should be performed. It was shown above how one could obtain values for all four of the quantities χ, η, I and A from a knowledge of any two of them. These four global properties of a chemical system reflect a change in the number of electrons and contain a wealth of information. The Fukui function $\partial\varrho/\partial N$, defined by Parr and Yang[64, 65], provides the necessary local property of the system. These five quantities, taken as a whole, will prove very useful in describing molecular properties and dynamics.

VI. References

1. Parr, R. G., Donnelly, R. A., Levy, M., Palke, W. E.: J. Chem. Phys. *68*, 3801 (1978)
2. Hohenberg, P., Kohn, W.: Phys. Rev. *136*, B864 (1964)
3. Levy, M.: Proc. Natl. Acad. Sci. USA *76*, 6062 (1979)
4. Lieb, E. H.: in: Physics as Natural Philosophy (Shimony, A., Feshbach, H., Eds.), Cambridge, Ma.: MIT Press 1982, p. 111
5. Levy, M.: Phys. Rev. A*26*, 1200 (1982)
6. Lieb, E. H.: Int. J. Quant. Chem. XXIV, 243 (1983)
7. Sanderson, R.T.: Science *114*, 670 (1951); see also Sanderson, R. T.: Chemical Bonds and Bond Energy, New York: Academic 1976
8. Donnelly, R. A., Parr, R. G.: J. Chem. Phys. *69*, 4431 (1978)
9. Mulliken, R. S.: ibid. *2*, 782 (1934)
10. Mulliken, R. S.: ibid. *3*, 573 (1935)
11. Pritchard, H. O., Summer, F. H.: Proc. R. Soc. London Ser. A*235*, 136 (1956)
12. Iczkowski, R. P., Margrave, J. L.: J. Am. Chem. Soc. *83*, 3547 (1961)
13. Hinze, J., Whitehead, M. A., Jaffe, H. H.: ibid. *85*, 148 (1963)

14. Baird, N. C., Sichel, J. M., Whitehead, M. A.: Theor. Chim. Acta *11*, 38 (1968)
15. Klopman, G.: J. Am. Chem. Soc. *86*, 1463 (1964)
16. Klopman, G.: ibid. *86*, 4450 (1964)
17. Klopman, G.: ibid. *87*, 3000 (1965)
18. Klopman, G.: J. Chem. Phys. *43*, 580 (1965)
19. Ferreira, R.: Adv. Chem. Phys. *13*, 55 (1967)
20. Fliszar, S.: J. Chem. Phys. *69*, 237 (1978)
21. Gyftopoulos, E. P., Hatsopoulos, G. N.: Proc. Natl. Acad. Sci. USA *60*, 786 (1968)
22. Perdew, J., Parr, R. G., Levy, M., Balduz, J. L.: Phys. Rev. Lett. *49*, 1691 (1982)
23. Perdew, J.: in Density Functional Methods in Physics (Dreizler, R. M., da Providencia, J., Eds.) New York: Plenum 1984
24. Parr, R. G., Bartolotti, L. J.: J. Phys. Chem. *87*, 2810 (1983)
25. Kohn, W., Sham, L.: Phys. Rev. *140*, A1133 (1965)
26. Levy, M.: Bull. Am. Phys. Soc. *24* (4), abstract EI 16 (1979)
27. Perdew, J., Zunger, A.: Phys. Rev. B*23*, 5048 (1981)
28. Slater, J. C.: The Self-Consistent Field for Molecules and Solids, New York: McGraw-Hill 1974
29. Janak, J. F.: Phys. Rev. B*18*, 7165 (1978)
30. Slater, J. C.: Int. J. Quantum Chem. *S3*, 727 (1970)
31. Johnson, K. H.: ibid. *S11*, 39 (1977)
32. Rajagopal, A. K.: Adv. Chem. Phys. XLI, 59 (1980)
33. Bamzai, A. S.: Deb, B. M.: Rev. Mod. Phys. *53*, 95 (1981)
34. March, N. H.: Theoretical Chemistry *4*, 92 (1981)
35. Parr, R. G.: Ann. Rev. Phys. Chem. *34*, 631 (1983)
36. Kohn, W., Vashishta, F.: in: Theory of the Inhomogeneous Electron Gas (Lundqvist, S., March, N. H., Eds.) New York: Plenum 1983, p. 79
37. Robles, J., Bartolotti, L. J.: J. Am. Chem. Soc. *106*, 3723 (1984)
38. Bartolotti, L. J., Gadre, S. R., Parr, R. G.: ibid. *102*, 2945 (1980)
39. Slater, J. C., Wood, J. H.: Int. J. Quantum Chem. *4S*, 3 (1971)
40. Cole, L. A., Perdew, J. P.: Phys. Rev. A*25*, 1265 (1982)
41. Pearson, R. G.: J. Am. Chem. Soc. *85*, 3533 (1963)
42. Parr, R. G., Pearson, R. G.: ibid. *105*, 7512 (1983)
43. Gopinathan, M. S., Whitehead, M. A.: Isr. J. Chem. *19*, 209 (1980)
44. Sen, K. D., Schmidt, P. C., Weiss, A.: Theor. Chim. Acta *58*, L69 (1980)
45. Sen, K. D.: J. Phys. B.: At. Mol. Phys. *16*, L149 (1983)
46. Schmidt, P. C., Böhm, M. C.: Bunsen Gesellschaft für Berichte der Physikalischen Chemie *87*, 925 (1983)
47. Manoli, S., Whitehead, M. A.: J. Chem. Phys. *81*, 841 (1984)
48. Gazquez, J. L., Ortiz, E.: ibid. *81*, 2741 (1984)
49. Sen, K. D., Schmidt, P. C., Weiss, A.: ibid. *75*, 1037 (1981)
50. Schwartz, K.: Phys. Rev. B*5*, 2466 (1972)
51. Schwartz, K.: Theor. Chim. Acta *34*, 225 (1974)
52. Moore, C. E.: Atomic Energy Levels, Natl. Bur. Stand, (US) Circ. No. 467 Vols. I–III
53. Hotop, H., Lineberger, W. C.: J. Phys. Chem. Ref. Data, Suppl. *4*, 539 (1975)
54. Handbbok on the Physics and Chemistry of Rare Earths (Gaschneider, Eyring, Eds.) Amsterdam: North-Holland Publishing Co. 1978, Vol. 1
55. Mulliken, R. S.: J. Chem. Phys. *46*, 497 (1948)
56. Moffit, W.: Proc. Roy. Soc. London Ser. A*202*, 531 (1950)
57. Moffit, W.: ibid. A*202*, 548 (1950)
58. Donnelly, R. A.: J. Chem. Phys. *71*, 2874 (1979)
59. Gunnarsson, O., Lundqvist, B. I.: Phys. Rev. B*13*, 4274 (1976)
60. Ray, N. K., Samuels, L., Parr, R. G.: J. Chem. Phys. *70*, 3680 (1979)
61. See for instance, Huheey, J. E.: Inorganic Chemistry: Principles of Structure and Reactivity, New York: Harper and Row 1972
62. Ceperely, D. M., Alder, B. J.: Phys. Rev. Lett. *45*, 566 (1980)
63. Bartolotti, L. J.: J. Chem. Phys. *76*, 6057 (1982)
64. Parr, R. G., Yang, W.: J. Am. Chem. Soc. *106*, 4049 (1984)
65. Yang, W., Parr, R. G., Pucci, R.: J. Chem. Phys. *81*, 2862 (1984)

Simple Density Functional Theory of the Electronegativity and Other Related Properties of Atoms and Ions

J. A. Alonso and L. C. Balbás

Departamento de Física Teórica, Universidad de Valladolid, Valladolid, Spain

Density functional theory (DFT) provides a convenient framework for quantitative computations of the electronegativity χ of atoms and ions. χ is just the negative of the chemical potential μ of DFT. In this paper we study χ using simple versions of the DFT within the original spirit of DFT, that is, working with the electron density $\varrho(\vec{r})$ as the fundamental ingredient. We show that density gradient terms must be taken into account to obtain results of qualitative or semiquantitative value. Apart from the study of the electronegativitiy of atoms and ions corresponding to several groups of the Periodic Table, the virtues of using DFT are illustrated by a number of selected applications: a) The electronegativity of exotic atoms containing unsaturated quarks attached to the nucleus. b) The consequences of electronegativity equalization for charge transfer in molecule formation. The relations between electronegativity, the electrostatic potential in the atom and bond distances are also explored following ideas originally put forward by Politzer. c) Finally, and due to its importance in catalytic processes, we make a start on the study of the electronegativity of small metallic particles.

1 Introduction 42

2 Simple Density Functional Theories of Atomic Electronegativity 44
- 2.1 Thomas-Fermi and Thomas-Fermi-Dirac Models 44
- 2.2 Improved Density Functionals. Weizsäcker-Type Models 45
- 2.3 Euler Equation 48

3 Results for the Weizsäcker-Type Model 49
- 3.1 Free Atoms and Ions 49
- 3.2 Quark Atoms 57

4 Applications to Molecules 60
- 4.1 Equalization of Electronegativities and Charge Transfer in Molecule Formation 60
- 4.2 Simple Charge Transfer Model of X-Ray Scattering by Ten-Electron Molecules . . . 65

5 Other Applications 69
- 5.1 Relation of Chemical Potential to Electrostatic Potential and Bonding Distances 69
- 5.2 Electronegativity, Ionization Potential and Electron Affinity of Metallic Microparticles 74

6 Summary 75

7 References 76

1 Introduction

Electronegativity is the power of an atom to attract electrons to itself in a molecule or solid compound[1]. This concept is then a convenient tool to deal with the ionic contribution to the chemical binding. If E(Z, N) is the total energy of an ion with nuclear charge +Ze as a function of the number N of electrons, the present theoretical view of the electronegativity of the ion is[2, 3]

$$\chi(Z, N) = -\frac{\partial E(Z, N)}{\partial N} . \tag{1}$$

The electronegativity of the neutral atom is then

$$\chi(Z, Z) = -\left[\frac{\partial E(Z, N)}{\partial N}\right]_{N=Z} . \tag{2}$$

The main difficulty with these equations is to extend E(Z, N) to a non-integral number of electrons[4, 5]. A method frequently used[6, 7] consists in writing E(Z, N) as a parabolic function of (N – Z) with coefficients determined from the experimental first ionization potential I_1 and electron affinity A of the atom, that is,

$$E(Z, N) = \frac{1}{2}(I_1 - A)(N - Z)^2 - \frac{1}{2}(I_1 + A)(N - Z) + E(Z, Z) . \tag{3}$$

Using this approximation in (2) one obtains

$$\chi^M(Z, Z) = \frac{I_1 + A}{2} . \tag{4}$$

which is the expression for the electronegativity of an atom proposed earlier by Mulliken[8]. Since

$$I_1 = E(Z, Z - 1) - E(Z, Z) , \tag{5}$$

$$A = E(Z, Z) - E(Z, Z + 1) , \tag{6}$$

$\chi^M(Z, Z)$ can be written

$$\chi^M(Z, Z) = \frac{E(Z, Z - 1) - E(Z, Z + 1)}{2} . \tag{7}$$

This result shows that $\chi^M(Z, Z)$ can be viewed as a finite difference approximation to $\chi(Z, Z)$. An interesting interpretation of $\chi^M(Z, Z)$ has been recently proposed by Komorowski[9]. This author has noticed that $\chi^M(Z, Z)$ is the average of $\chi(Z, N)$ in the interval (N = Z – 1, N = Z + 1):

$$\frac{\int_{Z-1}^{Z+1} \chi(Z, N)dN}{\int_{Z-1}^{Z+1} dN} = \frac{E(Z, N)\Big|_{N=Z-1}^{N=Z+1}}{N\Big|_{Z-1}^{Z+1}} = \frac{I_1 + A}{2} = \chi^M(Z, Z) . \tag{8}$$

Evidently, no explicit knowlegde of the functional form of E(Z, N) is needed to derive this result. The generalization of (7) to the case of a positive ion reads

$$\chi^M(Z, N) = \frac{1}{2}\left[E(Z, N-1) - E(Z, N+1)\right] . \tag{9}$$

One the other hand, the density functional (DF) formalism[10, 11] provides a natural and convenient framework to study the electronegativity. This formalism is based on the theorem, proved by Hohenberg and Kohn[10], which states that the ground state energy of a system of electrons in an external potential $V(\bar{r})$ is a functional $E_V[\varrho]$ of the electron density $\varrho(\bar{r})$. Minimization of this functional, subject to the normalization of the electron density leads to

$$\delta(E_V[\varrho] - \mu \int \varrho(\bar{r})\, d\bar{r}) = 0 . \tag{10}$$

If our system is an atom or an ion, this equation becomes

$$\delta(E[Z, N; \varrho] - \mu \int \varrho(\bar{r})\, d\bar{r}) = 0 . \tag{11}$$

which is equivalent to

$$\mu = \frac{\delta E[Z, N; \varrho]}{\delta \varrho(\bar{r})} . \tag{12}$$

The Lagrange multiplier μ, which is independent of $\bar{r}$, is usually called the chemical potential of the electrons. From (11) one can also write

$$\mu = \frac{\partial E(Z, N)}{\partial N} . \tag{13}$$

A comparison between (13) and (1) shows that the electronegativity χ is given by the negative of the chemical potential μ[11]

$$\chi = -\mu . \tag{14}$$

All that remains to be done to calculate the electronegativity is to write down the explicit form of the energy functional $E[Z, N; \varrho]$ and then solve the Euler equation (12). This is, in fact, a difficult task since the exact form of the functional $E_V[\varrho]$ is unknown, and one must resort to approximations.

The objective of this work is to obtain χ using simple approximations to $E_V[\varrho]$, working within the original spirit of the density functional formalism, that is, with the electron density $\varrho(\bar{r})$ as the basic ingredient. More refined calculations within the DF formalism exist which are based on one-electron wave functions Ψ_i (such that the electron density is built as $\varrho = \sum_i |\Psi_i|^2$), but those will only be referred briefly for comparative purposes. What we try to show is what level of sophistication is needed in the energy functional, expressed in terms of the electron density, to obtain results of qualitative or semiquantitative value. After presenting approximate DF theories and their results for atoms and ions in Sects. 2 and 3, some applications to chemical binding in molecules are offered in Sect. 4. Those applications are based on the principle of equalization of electronegativities due to Sanderson[12, 13]. This principle states that the electronegativities of the component atoms become equalized in the molecule. Finally Sect. 5 deals with applications to other systems.

2 Simple Density Functional Theories of the Atomic Electronegativity

2.1 Thomas-Fermi and Thomas-Fermi-Dirac Models

The ground state energy of an atom or ion can be written as a functional of the electron density[10] (Hartree atomic units are used in the paper unless otherwise stated)

$$E[Z, N; \varrho] = -\int \frac{Z}{r} \varrho(\bar{r})\, d\bar{r} + \frac{1}{2} \int\int \frac{\varrho(\bar{r})\, \varrho(\bar{r}')}{|\bar{r} - \bar{r}'|}\, d\bar{r}\, d\bar{r}' + F[\varrho]. \tag{15}$$

The first term is the interaction of the electron with the nuclear potential, the second term is the classical coulombic interaction between the electrons and the third term is a functional which contains all the remaining contributions to $E[Z, N; \varrho]$. These contributions are the kinetic energy $T[\varrho]$ and the exchange and correlation energies, $E_x[\varrho]$ and $E_c[\varrho]$ respectively.

The simplest density functional theory is the Thomas-Fermi (TF) model[14, 15]. In this model exchange and correlation effects are neglected and the kinetic energy is approximated locally (that is, at point $\bar{r}$) by the kinetic energy density of a free electron gas with (constant) density equal to the density $\varrho(\bar{r})$ at point $\bar{r}$. This leads to the following approximation to $F[\varrho]$:

$$F_{TF}[\varrho] = T_{TF}[\varrho] = \int c_k \varrho(\bar{r})^{5/3} d\bar{r}\ , \tag{16}$$

with $c_k = \frac{3}{10}(3\pi^2)^{2/3}$. Using (16) in (15) and then solving the Euler equation (12) with boundary conditions appropriate to a free atom, leads to the well known result[16]

$$\mu(TF) = 0\ . \tag{17}$$

This means that all neutral atoms have the same electronegativity in the TF model and that no driving force exists for the transfer of electronic charge between atoms in a molecule. This simple model is not useful in dealing with χ.

The next level of sophistication consists in adding exchange, also in a local approximation[16]. Then

$$F_{TFD}[\varrho] = \int c_k\, \varrho(\bar{r})^{5/3} d\bar{r} + \int c_x \varrho(\bar{r})^{4/3} d\bar{r} \,. \tag{18}$$

where $c_x = -3/4(3/\pi)^{1/3}$. The resulting model is the Thomas-Fermi-Dirac (TDF) model[16]. Using (18) in (15) and solving the corresponding Euler equation one arrives at the result[17)]

$$\mu(\text{TDF}) = \text{constant} = \frac{1}{2\pi^2} \,. \tag{19}$$

for the solutions which can be interpreted to correspond to free neutral atoms. Another way of viewing Eqs. (17) and (19) is through Teller's theorem[18, 19] which states that chemical binding between atoms does not exist in the TF or TFD models (that is, for a theory with purely local density contributions to the electronic kinetic energy).

2.2 *Improved Density Functionals. Weizsäcker-Type Models*

Since a purely local density approximation ot $T[\varrho]$ is not capable of giving qualitatively correct results for χ, the use of more accurate functionals is necessary. We now introduce an improved approximation to $T[\varrho]$ which has been used by us in the study of the electronegativity[20–26]. The starting point is the expression for the kinetic energy density $t_{\bar{r}}$ at point $\bar{r}$, written in terms of the first order density matrix[27, 28]

$$t = t_{\bar{r}} = \frac{1}{2}\Big[\nabla \cdot \nabla' \gamma(\bar{r}, \bar{r}')\Big]_{\bar{r}=\bar{r}'} \,. \tag{20}$$

$t_{\bar{r}}$ is such that $\int t_{\bar{r}} d\bar{r} = T$, where T is the correct total kinetic energy of the system. From Eq. (20) and following the derivation of Alonso and Girifalco[29] and others[30–33] one arrives at the following result, valid for atoms with doubly occupied orbitals in the restricted Hartree-Fock (HF) approximation

$$t_{\bar{r}}^{HF} = \frac{1}{8}\frac{(\nabla\varrho(\bar{r}))^2}{\varrho(\bar{r})} - \frac{1}{2}\,\varrho(\bar{r})\Big[\nabla \cdot \nabla' C(\bar{r}, \bar{r}')\Big]_{\bar{r}=\bar{r}'} \,. \tag{21}$$

The first term in (21) is usually known as the Weizsäcker term $t^W[\varrho]$[34]. It has been recently emphasized by several authors that this term is a natural component of the exact ground state kinetic energy density functional[29–33, 35–37]. In particular, this term gives the exact ground state kinetic energy of a one-electron atom (ion), or of a two electron atom (ion) in the Hartree-Fock approximation. $C(\bar{r}, \bar{r}')$ is the statistical Pauli-exchange hole around an electron placed at the position $\bar{r}$. A simple approximation for $C(\bar{r}, \bar{r}')$ is to use the Hartree-Fock form for a homogeneous non-polarized system of constant density ϱ_h

$$C_{HF}(\bar{r}, \bar{r}') = -\frac{9}{2}\left(\frac{\sin x - x\cos x}{x^3}\right)^2 \tag{22}$$

$$x = (3\pi^2\varrho_h)^{1/3} \cdot |\bar{r} - \bar{r}'| , \tag{23}$$

plus the local (L) density approximation $\varrho_h = \varrho(\bar{r})$. With this result in (21), $t_{\bar{r}}^{HF}$ can be approximated by the following functional of ϱ[29]

$$t_{\bar{r}}^{HFL} = t_{\bar{r}}^{HFL}[\varrho] = \frac{1}{8}\frac{(\nabla\varrho(\bar{r}))^2}{\varrho(\bar{r})} + c_k\varrho(\bar{r})^{5/3} . \tag{24}$$

A virtue of this functional is that it gives the correct asymptotic behaviour of $t_{\bar{r}}$ for $r \to \infty$. Nevertheless, if accurate (HF or better) electron densities are used in (24), it is found that the total kinetic energy T of atoms is badly overestimated. There are some ways to correct for this fact. Acharya and coworkers[36] have proposed an improved form of this functional, namely

$$t_{\bar{r}}^{A}[\varrho] = \frac{1}{8}\frac{(\nabla\varrho(\bar{r}))^2}{\varrho(\bar{r})} + \left(1 - \frac{B}{N}\right)^{1/3} c_k\varrho(\bar{r})^{5/3} , \tag{25}$$

where N is the number of electrons of the atom and B is a constant, obtained by a global fit using HF data (for $\varrho(\bar{r})$ and T) for 1200 atoms and ions. The form of the factor $\left(1 - \frac{B}{N}\right)^{1/3}$ has been justified by further work of Gázquez and Robles[31]. The functional of Eq. (25) has been studied by Bartolotti and Acharya[38]. The main conclusion of these workers is the necessity of modelling $(t - t^W)$ with functionals which are nonlocal in the density.

An improved description of the kinetic energy is obtained separating the electrons of the atom in groups (for instance, core (c) and valence (v) electrons) and assuming that the first order density matrix can be written as a sum over groups of group density matrices

$$\gamma(\bar{r}, \bar{r}') = \gamma_c(\bar{r}, \bar{r}') + \gamma_v(\bar{r}, \bar{r}') . \tag{26}$$

Repeating again the arguments of Alonso and Girifalco[29] and others[30–33] within each group one arrives at a generalization of Eq. (21)

$$\begin{aligned} t_{\bar{r}}^{HF} = &\frac{1}{8}\frac{(\nabla\varrho_c(\bar{r}))^2}{\varrho_c(\bar{r})} - \frac{1}{2}\varrho_c(\bar{r})\left[\nabla \cdot \nabla' C_c(\bar{r}, \bar{r}')\right]_{\bar{r}'=\bar{r}} \\ &+ \frac{1}{8}\frac{(\nabla\varrho_v(\bar{r}))^2}{\varrho_v(\bar{r})} - \frac{1}{2}\varrho_v(\bar{r})\left[\nabla \cdot \nabla' C_v(\bar{r}, \bar{r}')\right]_{\bar{r}'=\bar{r}} , \end{aligned} \tag{27}$$

where ϱ_c and ϱ_v are the core and valence electron densities respectively, and the meaning of C_c (C_v) is that only core (valence) electrons contribute to the exchange hole around a core (valence) electron. A local HF approximation to C_c and C_v (see Eqs. 22, 23) then leads to

$$t_{\bar{r}}^{HFL}[\varrho] = \frac{1}{8}\frac{(\nabla\varrho_c(\bar{r}))^2}{\varrho_c(\bar{r})} + c_k\varrho_c(\bar{r})^{5/3} + \frac{1}{8}\frac{(\nabla\varrho_v(\bar{r}))^2}{\varrho_v(\bar{r})} + c_k\varrho_v(\bar{r})^{5/3} . \tag{28}$$

In the results to be presented below the frozen-core approximation will be made, the effect of the core will be simulated by a local pseudopotential, and only valence electrons will be taken into account. In this case, the kinetic energy (of the valence electrons) is

$$T^{HFL}[\varrho_v] = \int t_{\bar{r}}^{HFL}[\varrho_v]\,d\bar{r} = \int \left\{ \frac{1}{8} \frac{(\nabla \varrho_v(\bar{r}))^2}{\varrho_v(\bar{r})} + c_k \varrho_v(\bar{r})^{5/3} \right\} d\bar{r} \,, \tag{29}$$

which is one of the basic ingredients that we use in the computations to be described later.

We nevertheless point out that it is possible to go beyond Eq. (28) or (24) using a non-local weighted density HF approximation to $C(\bar{r}, \bar{r}')$ in Eqs. (27) or (21). Let us again take the HF form for $C(\bar{r}, \bar{r}')$, given in Eqs. (22) and (23), but with a density parameter $\varrho_h = \bar{\varrho}(\bar{r})$, where $\bar{\varrho}(\bar{r})$ is a weighted density (WD) calculated from the normalization condition of the exchange-charge around an electron placed at $\bar{r}$, that is

$$\int \varrho(\bar{r}')\, C_{HF}(\bar{r}, \bar{r}'; \bar{\varrho}(\bar{r}))\,d\bar{r}' = -1 \,. \tag{30}$$

Then the corresponding kinetic energy functional becomes (restricting our consideration to the valence-only case)

$$T^{WD}[\varrho_v] = \int t_{\bar{r}}^{WD}[\varrho_v]\,d\bar{r} = \int \left\{ \frac{1}{8} \frac{(\nabla \varrho_v(\bar{r}))^2}{\varrho_v(\bar{r})} + c_k \varrho_v(\bar{r}) \bar{\varrho}_v(\bar{r})^{2/3} \right\} d\bar{r} \,. \tag{31}$$

A study of the Weizsäcker-type functionals for the kinetic energy of Hartree-Fock atoms has been presented elsewhere[39]. Of course (31) is more accurate than (29), although it is much more difficult to solve the Euler equation associated with (31)[40].

The energy functional (for the valence electrons) to be used in our study of electronegativities is then

$$E[V_I, N_v; \varrho_v] = \int V_I(\bar{r}) \varrho_v(\bar{r})\,d\bar{r} + \frac{1}{2} \int\int \frac{\varrho_v(\bar{r}) \varrho_v(\bar{r}')}{|\bar{r} - \bar{r}'|}\,d\bar{r}\,d\bar{r}' + T^{HFL}[\varrho_v] + E_x[\varrho_v] + E_c[\varrho_v] \,. \tag{32}$$

In this equation $V_I(\bar{r})$ is the ionic pseudopotential, N_v is the number of valence electrons of the atom or ion, $T^{HFL}[\varrho_v]$ is the kinetic energy functional of Eq. (21), $E_x[\varrho_v]$ is the local-density exchange energy functional [the second term in Eq. (18)] and $E_c[\varrho_v]$ is the local density approximation to the correlation energy functional, constructed from Wigner's formula for a homogeneous electron system[41)]

$$E_c[\varrho_v] = - \int \frac{0.44\, \varrho_v(\bar{r})}{7.8 + (4\pi \varrho_v(\bar{r})/3)^{-1/3})}\,d\bar{r} \,. \tag{33}$$

Finally, the empty-core model pseudopotential will be taken for V_I[42)]

$$V_I(\bar{r}) = \begin{cases} 0 \,, & r < R_c \\ -\dfrac{Z_I}{r} \,, & r > R_c \end{cases} \tag{34}$$

where R_c is the empty core radius and Z_I is the positive charge of the ion core.

2.3 Euler Equation

Minimization of $E[V_I, N_v; \varrho_v]$ subject to normalization of the valence electron density leads to the Euler equation

$$\mu = V_T(\bar{r}) + \frac{5}{3} c_k \varrho_v(\bar{r})^{2/3} + \frac{1}{8}\left(\frac{\nabla \varrho_v(\bar{r})}{\varrho_v(\bar{r})}\right)^2$$

$$-\frac{1}{4}\frac{\nabla^2 \varrho_v(\bar{r})}{\varrho_v(\bar{r})} - \frac{4}{3} c_x \varrho_v(\bar{r})^{1/3} - \frac{C_1 + C_2 \varrho_v(\bar{r})^{-1/3}}{[7.8 + (4\pi \varrho_v(\bar{r})/3)^{-1/3}]^2}, \tag{35}$$

where $C_1 = 3.43$, $C_2 = 0.36$ and $V_T(\bar{r})$ is the total electrostatic potential

$$V_T(\bar{r}) = V_I(\bar{r}) + V_e(\bar{r}) = V_I(\bar{r}) + \int \frac{\varrho_v(\bar{r}')}{|\bar{r} - \bar{r}'|} d\bar{r}' . \tag{36}$$

Equation (35) must be now integrated, subject to the selfconsistency condition imposed by Poisson's equation

$$\Delta V_T(\bar{r}) = \Delta(V_I(\bar{r}) + V_e(\bar{r})) = -4\pi \varrho_v(\bar{r}) , \tag{37}$$

and subject also to appropriate boundary conditions. The integration is started at the center of the atom with initial values

$$\left.\frac{d\varrho_v}{dr}\right|_{r=0} = \left.\frac{dV_T}{dr}\right|_{r=0} = 0$$

$$\varrho_v(0) = \varrho_0 ; \quad V_T(0) = V_0 , \tag{38}$$

where ϱ_0 and V_0 are arbitrary constants. Then ϱ_0 and V_0 are varied until the following conditions are simultaneously satisfied in the asymptotic region of a normalized solution

$$\varrho_v(\bar{r}) \xrightarrow[r \to \infty]{} 0 ; \quad \frac{d\varrho_v}{dr} \xrightarrow[r \to \infty]{} 0 . \tag{39}$$

After the electron density has been obtained, the energy is readily computed from the functional $E[V_I, N_v; \varrho_v]$. The electronegativity $\chi = -\mu$ is, of course, a natural outcome of the calculation. Evaluating (35) for the limit $r \to \infty$, it is easy to show that the electron density decays exponentially in the asymptotic region and that this exponential decay is controlled by μ[43, 44].

The results of selfconsistent calculations using Eqs. (35) and (37) will be presented in Sect. 3 below. But a preliminary idea can be gathered if accurate electron densities taken from an independent source are directly used in (35) to evaluate μ. In such a case μ becomes a function of $\bar{r}$. With this shortcoming in mind, Eq. (35) can be evaluated in the limit $r \to \infty$[20], this limit being chosen because, as stated above, the functional of Eq. (24) is asymptotically correct for $r \to \infty$. The asymptotic behaviour of the atomic electron density has been investigated by several authors[45–48]. According to them, in the asymptotic region

$$\varrho(\bar{r}) \sim r^{\beta} e^{-[2(2I_1)^{1/2}]} . \tag{40}$$

Using this result in (35) one obtains

$$\chi(Z, Z) = -\mu(r \to \infty) = I_1 \tag{41}$$

which is a result qualitatively correct.

3 Results for the Weizsäcker-Type Model

3.1 Free Atoms and Ions

The only parameter in the approximate density functional theory developed above is the empty-core radius R_c. For each chemical element, R_c has been fixed by requiring that the calculated energy of the valence shell of the neutral atom equals the experimental energy, that is, the sum of the measured ionization potentials of the electrons in the valence shell[49]. For the Li, Be, B, C, N, O and F groups the number of valence electrons is 1, 2, 3, 4, 5, 6 and 7 respectively. The core radii calculated in this way for atoms of these groups are given in Table 1. Calculated ionization potentials are given in Figs. 1 and 2. These have been obtained by subtracting calculated total energies, that is,

$$I_1 = E(Z, N = Z - 1) - E(Z, N = Z) \tag{42}$$

$$I_2 = E(Z, N = Z - 2) - E(Z, N = Z - 1) , \tag{43}$$

etc. The experimental trends are well reproduced by the calculated ionization potentials.

Table 1. Calculated core radii (atomic units)

Atom	R_c	Atom	R_c	Atom	R_c	Atom	R_c
Li	1.094	Be	0.646	B	0.519	C	0.399
Na	1.237	Mg	0.996	Al	0.968	Si	0.861
K	1.824	Ca	1.572	Ga	0.839	Ge	0.867
Rb	1.972	Sr	1.806	In	0.988	Sn	1.031
Cs	2.274	Ba	2.107	Tl	0.868	Pb	0.966
Atom	**R_c**	**Atom**	**R_c**	**Atom**	**R_c**		
N	0.312	O	0.255	F	0.205		
P	0.759	S	0.676	Cl	0.595		
As	0.822	Se	0.775	Br	0.728		
Sb	1.014	Te	0.943				
Bi	1.010						

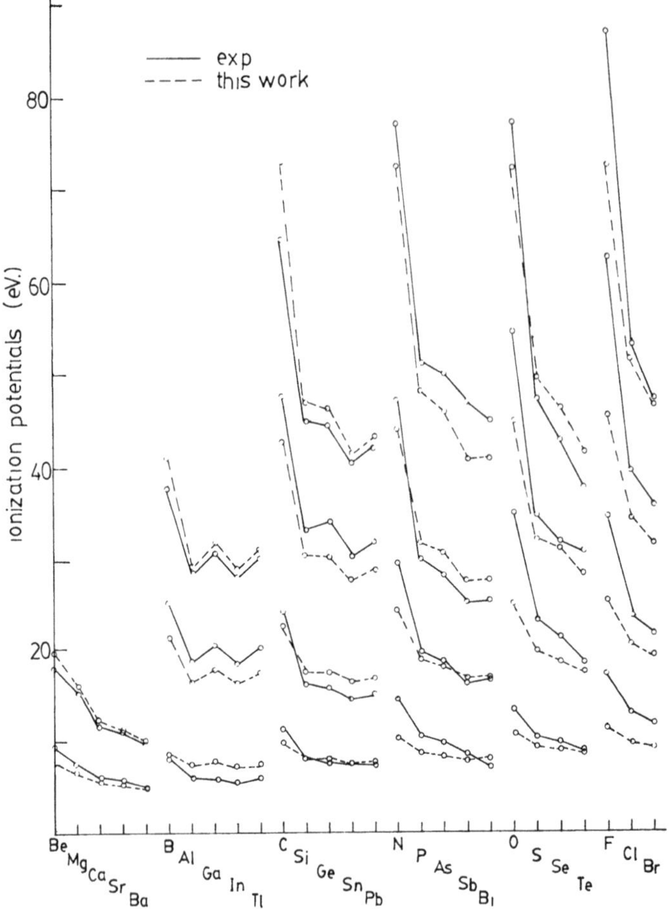

Fig. 1. Comparison of calculated and experimental first (I_1), second (I_2), third (I_3) and fourth (I_4) ionization potentials of atoms of the groups IIA to VIIA of the Periodic Table. Only I_1 and I_2 are given for the elements of group IIA. I_1, I_2 and I_3 are given for the elements of group IIIA and I_1, I_2, I_3 and I_4 for groups IVA to VIIA. To identify each ionization potential, notice that $I_1 < I_2 < I_3 < I_4$

Nevertheless, we stress from the beginning that the approximate density functional method used in this section is not intended to compete with accurate quantum mechanical methods. Our main purpose is not to compute ionization potentials but to study the electronegativity of atoms and ions using a method as simple as possible. Since the electronegativity is closely related to ionization potentials we then expect that the approximate density functional method can be used with some confidence in the study of the electronegativity. The calculated electronegativities of neutral atoms $\chi_{DF}(Z, Z)$ are listed in Table 2 and those of positive ions are plotted in Figs. 3 and 4. In these two figures we have also plotted Mulliken electronegativities (see Eq. (9)), computed in two different ways. The set χ^M_{ex} was obtained from experimental energies $E(Z, N \pm 1)$, whereas calculated energies were used to obtain χ^M_{DF}. The three quantities show the same trends for the set of ions studied. χ_{DF} is slightly smaller than χ^M_{DF}. This is due to the difference between the true derivative in (1) and its finite difference approximation,

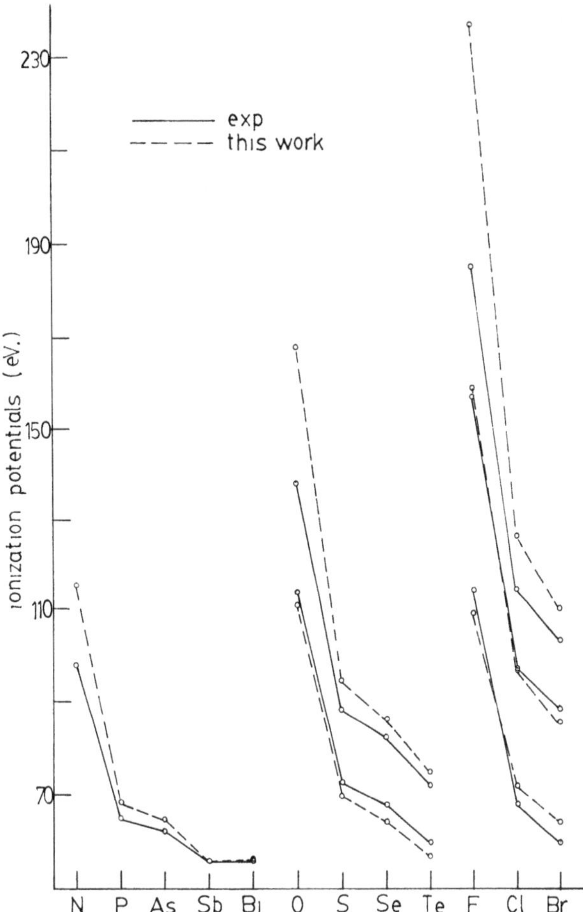

Fig. 2. Comparison of calculated and experimental ionization potentials of atoms of the groups VA, VIA and VIIA of the Periodic Table. Only I_5 is given for group VA. I_5 and I_6 are given for group VIA and I_5, I_6, I_7 for group VIIA. To identify each ionization potential, notice that $I_5 < I_6 < I_7$

given in Eq. (9). The systematics of the difference $(\chi_{DF}^{M} - \chi_{DF})$ is apparent in Fig. 5. For a given chemical element, the fractional deviation decreases as the degree of ionization increases. A few exceptions to this rule are found for elements of low atomic number. Extrapolation of the results suggests significant deviations (perhaps of about 10%) between χ_{DF}^{M} and χ_{DF} for neutral atoms[23]. This fact should be kept in mind when comparing χ_{DF} with χ_{ex}^{M} for neutral atoms and this might be also true in more accurate theories. Period and group effects are clearly visible in Fig. 5.

Parr and Pearson[50] have recently defined the "absolute hardness" of an atom by

$$\eta = \frac{1}{2}\left[\frac{\partial^2 E(Z, N)}{\partial N^2}\right]_{N=Z} = -\frac{1}{2}\left[\frac{\partial \chi(Z, N)}{\partial N}\right]_{N=Z} = -\frac{1}{2}\chi' \tag{44}$$

and they have proposed that the absolute hardness is essentially equivalent to the chemi-

Table 2. Density functional electronegativity and hardness of neutral atoms (in eV)

Atom	χ	η	Atom	χ	η	Atom	χ	η
Li	2.317	2.572	Be	3.415	3.307	B	4.120	3.714
Na	2.216	2.600	Mg	3.187	2.929	Al	3.706	3.176
K	1.999	2.174	Ca	2.723	2.421	Ga	3.842	3.368
Rb	1.988	1.674	Sr	2.580	2.223	In	3.657	3.022
Cs	1.922	1.735	Ba	2.573	2.057	Tl	3.782	3.179
Atom	χ	η	Atom	χ	η	Atom	χ	η
C	4.680	4.064	N	5.133	4.345	O	5.404	4.562
Si	4.187	3.403	P	4.640	3.598	S	4.988	3.790
Ge	4.180	3.443	As	4.589	3.514	Se	4.900	3.632
Sn	4.057	3.257	Sb	4.398	3.276	Te	4.746	3.394
Pb	4.135	3.214	Bi	4.408	3.231			
Atom	χ	η						
F	5.606	4.679						
Cl	5.318	3.932						
Br	5.180	3.758						

cal hardness concept as empirically developed for chemical reactions[51, 52]. χ' has been computed by several authors using Eq. (3), which leads to

$$\eta = \tfrac{1}{2}(I_1 - A) \tag{45}$$

In Table 2 we give calculated density functional values of η, obtained by first writing

$$\chi(Z, N) = \chi(Z, Z) + a_1(N - Z) + a_2(N - Z)^2 \tag{46}$$

and then computing a_1 and a_2 by fitting Eq. (46) to calculated values of $\chi_{DF}(Z, Z - 0.2)$, $\chi_{DF}(Z, Z)$ and $\chi_{DF}(Z, Z + 0.2)$. These electronegativities correspond to the neutral atom and to fictitious free ions with net charges ± 0.2. The conditions of non-integer electron number is imposed through the normalization required to solve the Euler equation. Evidently from (46) one obtains.

$$\eta_{DF} = a_1 \, . \tag{47}$$

η_{DF} is in reasonable qualitative agreement with results obtained by other workers using (45)[50, 53], but in general η_{DF} is smaller than η of (45). This is, in our opinion, due to the fact that the theoretical I_1 is smaller than the experimental I_1 in most cases, as can be seen in Fig. 1 (see further comments about this point later). A fit of (46) to four calculated electronegativities $\chi_{DF}(Z, Z - 1)$, $\chi_{DF}(Z, Z - 0.2)$, $\chi_{DF}(Z, Z)$ and $\chi_{DF}(Z, Z + 0.2)$ produces values of η_{DF} very similar to those given in Table 2.

The density functional pseudopotential model given in Eqs. (32–34) achieves our main goal of going beyond the unphysical results of the TF or TFD models ($\mu_{TF}(Z, Z) =$

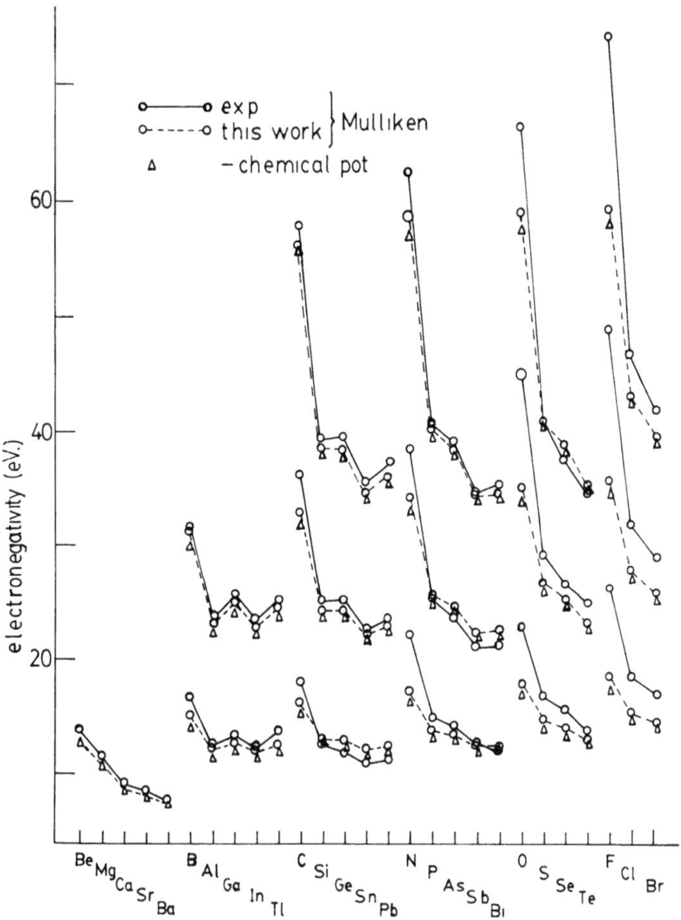

Fig. 3. Electronegativity of positive ions of elements X of the groups IIA to VIIA of the Periodic Table. Only $\chi(X^+)$ is given for elements of group IIA. $\chi(X^+)$ and $\chi(X^{2+})$ are given for elements of group IIIA. $\chi(X^+)$, $\chi(X^{2+})$ and $\chi(X^{3+})$ are given for elements of groups IVA to VIIA. To identify each ion, notice that $\chi(X^+) < \chi(X^{2+}) < \chi(X^{3+})$. Three values of χ are given for each ion: one (χ_{DF}) is the electronegativity obtained by solving the Euler equation of the DF theory. Mulliken's approximation has been used to obtain the other two values (one of them, χ^M_{ex}, using experimental energies in Eq. (9), and the other one, χ^M_{DF}, using calculated energies)

0 and $\mu_{TFD}(Z, Z) = 1/2\,\pi^2$ respectively), while still maintaining the simplicity of a theory formulated in terms of the electron density. But how accurate the calculated DF electronegativities are? Figure 6 compares the electronegativities of Table 2 with those calculated by Robles and Bartolotti[54]. These authors have employed the Kohn-Sham[55] formulation of density functional theory with a local exchange correlation functional[56]. There is rough agreement between the two sets for elements of valency up to four. On the other hand the electronegativities of Table 2 are systematically smaller for valencies larger than four, although a linear correlation, with slopes different form unity, also exists for this set of elements. Similar conclusions are obtained if comparison is made with the electronegativities computed by Bartolotti, Gadre and Parr[57], who used the Xα

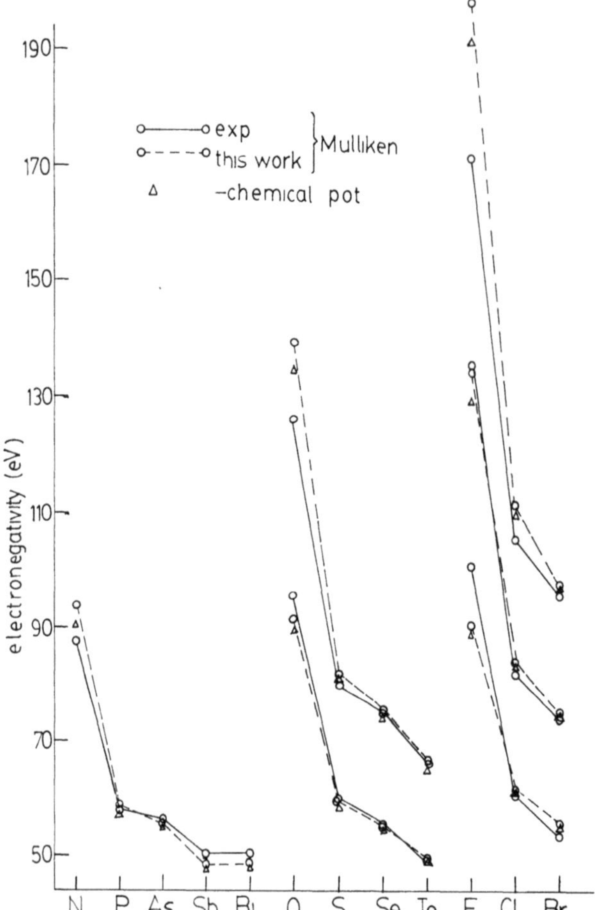

Fig. 4. Electronegativity of positive ions of elements X of the groups VA to VIIA of the Periodic Table. Only $\chi(X^{4+})$ is given for elements of the group VA. $\chi(X^{4+})$ and $\chi(X^{5+})$ are given for elements of the group VIA, and $\chi(X^{4+})$, $\chi(X^{5+})$, $\chi(X^{6+})$ for elements of the group VIIA. To identify each ion, notice that $\chi(X^{4+}) < \chi(X^{5+}) < \chi(X^{6+})$. Three values of χ are given for each ion (see caption of Fig. 3 for an explanation)

theory[58]. In conclusion, the calculated electronegativities χ_{DF} of neutral atoms show roughly the same trends as those calculated by more accurate methods, although the range of electronegativity values spanned is more reduced. This will affect quantitatively the trends in the ionic contribution to the chemical binding but not the qualitative trends.

Most of the points plotted in Fig. 6 are such that $\chi_{DF} < \chi_{RB}$. This fact can be corrected in part by choosing empty core radii smaller than those given in Table 1[21] but in this way only a part of the discrepancy is removed and there remains another part due to the intrinsic approximate nature of the functional $T^{HFL}[\varrho_v]$ of Eq. (29).

An interesting property of the chemical potential $\mu(Z, N)$ of an atomic ion has been derived by March[59] using Feynman's theorem, which yields[60, 61]

$$-V_e(Z, N; r = 0) = -\int \frac{\varrho(Z, N; \bar{r})}{r} d\bar{r} = \frac{\partial E(Z, N)}{\partial Z} . \tag{48}$$

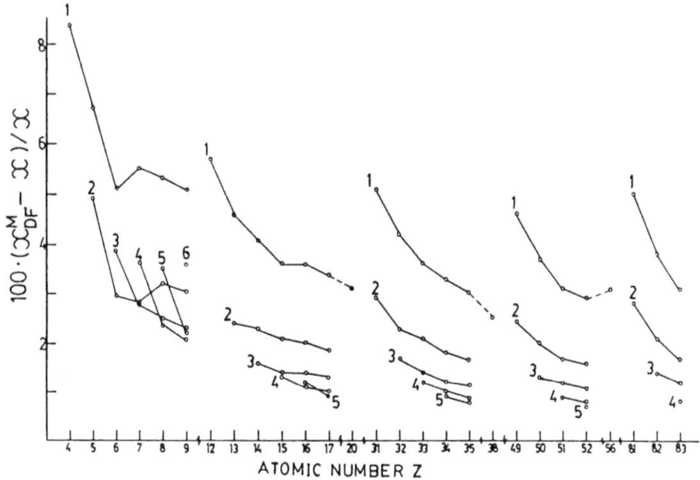

Fig. 5. Relative difference between the density functional electronegativity and its Mulliken counterpart (see Text) for positive ions. Elements are disposed in order of increasing atomic number. The degree of ionization is identified by a number between 1 and 6. Ions with the same net positive charge are joined by lines

Here $V_e(Z, N; r = 0)$ is the electrostatic potential of the electron cloud at the nucleus of the ion. Let us notice that

$$\frac{\partial \mu(Z, N)}{\partial Z} = \frac{\partial^2 E(Z, N)}{\partial N \, \partial Z} . \tag{49}$$

Differentiating now Eq. (48) partially with respect to N and comparing the result with (49), March obtained the exact relation

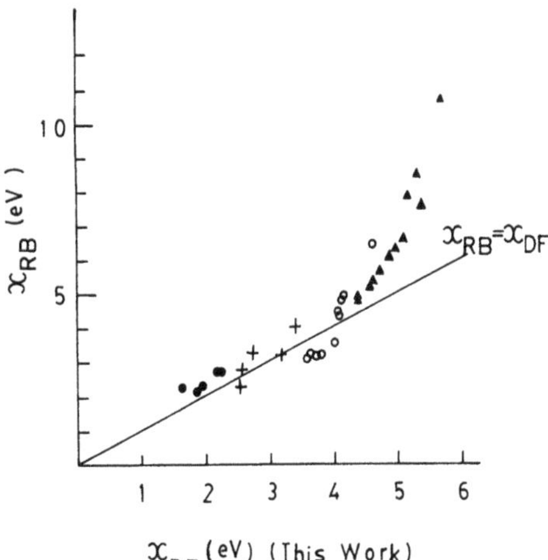

Fig. 6. Calculated DF electronegativity χ_{DF} versus Robles-Bartolotti electronegativity χ_{RB}. ● Valence 1; + Valence 2; ○ Valencies 3 and 4; ▲ Valence larger than 4

$$\frac{\partial\mu(Z, N)}{\partial Z} = \frac{\partial V_e(Z, N; r = 0)}{\partial N} . \tag{50}$$

Knowledge of $V_e(Z, N; r = 0)$ from accurate electron densities provides a route to $\partial\mu/\partial Z$ from Eq. (50), which would constitute valuable information about the chemical potential. To press this point further, March used the following relation, valid in TF theory of atomic ions[62)]

$$- ZV_e(Z, N; r = 0) = \frac{7}{3} E(Z, N) - N\mu(Z, N) . \tag{51}$$

(of course Eq. (17) is a particular case of this equation). Differentiating (51) partially with respect to N one obtains

$$- Z \frac{\partial V_e(Z, N; r = 0)}{\partial N} = \frac{4}{3} \mu(Z, N) - N \frac{\partial\mu(Z, N)}{\partial Z} . \tag{52}$$

Finally, using this result in (50), the following relation between the partial derivatives of μ with respect to Z and N is found

$$Z \frac{\partial\mu(Z, N)}{\partial Z} + \frac{\partial\mu(Z, N)}{\partial N} = \frac{4}{3}\mu(Z, N) . \tag{53}$$

This equation can also be viewed in another way. It is readily shown to be a consequence of the scaling of μ in TF theory, namely $\mu(Z, N) = Z^{4/3} f(N/Z)$[62)], or from the equivalent fact that $E_{TF}(Z, N) = Z^{7/3} F(N/Z)$ (f is the derivative of F with respect to its argument). Although Eq. (51) is a result of the simple TF model, it has been found to be approximately valid in Hartree-Fock theory[63–66)]. This fact suggests that it is worth investigating whether Eq. (53) is also approximately valid beyond the TF model. In other words, it should be desirable to test Eq. (53) using chemical potentials obtained from more accurate density functional theories.

From the results for $\mu(Z, N)$ obtained above for the Weizsäcker-type model (and given in Table 2 and Figs. 3 and 4), we calculate $\partial\mu/\partial Z$ as

$$\frac{\partial\mu(Z, N)}{\partial Z} = \tfrac{1}{2} [\mu(Z + 1, N) - \mu(Z - 1, N)] , \tag{54}$$

which is equivalent to a quadratic fit to the chemical potentials $\mu(Z + 1, N)$, $\mu(Z, N)$ and $\mu(Z - 1, N)$, corresponding to three consecutive members of the isoelectronic series with N electrons. In a similar way $\partial\mu/\partial N$ is obtained as

$$\frac{\partial\mu(Z, N)}{\partial N} = \tfrac{1}{2} [\mu(Z, N + 1) - \mu(Z, N - 1)] . \tag{55}$$

These two derivatives have been calculated for twenty-four atomic ions and the results were fitted[25)] to the equation

$$Z \frac{\partial \mu}{\partial Z} + BN \frac{\partial \mu}{\partial N} = A\mu . \tag{56}$$

The values of A and B obtained in the fit are A = 1.395, B = 1.028, and the correlation coefficient is r = 0.9996. These values are very close to the corresponding values in Eq. (53), that is, Eq. (53) is also approximately valid in other density functional models more sophisticated than the simple TF model.

Equation (53) was also tested, using HF theory to construct the various terms in the equation[25]. In this case μ is the least negative occupied one-electron energy[67]. Then, the derivatives of μ were again calculated from Eqs. (54) and (55) for twenty ions, taking the necessary HF eigenvalues from Clementi and Roetti[68]. A fit to Eq. (56) gave A = 1.72, B = 1.01 with a correlation coefficient r = 0.970. This result indicated that also in HF theory the chemical potential is approximately a homogeneous magnitude in both Z and N, although the degree of homogeneity (A = 1.72) is higher than in TF theory, which reflects the deviation of the kinetic energy from the local density (TF) approximation and also the effects of exchange. However a detailed analysis shows that the conclusion about the approximate homogeneity of μ in theories more sophisticated than the TF model must be interpreted only in an average sense. This means that separate fits of Eq. (56) in several isoelectronic series lead to slightly different values of A and B and that the values given above are average values. This occurs for both the Weizsäcker-type and the HF models[25].

3.2 Quark Atoms

Several works have recently discussed the existence of atoms with fractional nuclear charge[53, 69–79]. Such atoms are composed of a central nucleus of small size with positive fractional charge, surrounded by an electronic cloud of normal atomic dimensions[74]. The fractional charge of the nucleus is due to a free fractionally charged particle attached to the nucleus. This particle could be a quark. It has been proposed that these free quarks are survivers of the "Big Bang"[53, 74]. These atoms are sometimes known as quark atoms.

The interactions of quark atoms with ordinary atoms are governed by the usual laws of chemistry[53, 74]. The chemical properties of a quark atom depend only on the fractional charge of the nucleus and the number of electrons. An effective search for quark atoms in matter requires an understanding of the chemistry of quark atoms. A better understanding of this chemistry is helpful in the design and interpretation of quark search experiments and suggests host minerals where quark atoms might be found. Lackner and Zweig[53, 74] have studied the electronegativity of quark atoms using Mulliken's approximation [Eq. (4)], which can be written for this case

$$\chi^M(Z + \delta, Z) = \frac{I_1(\delta) + A(\delta)}{2} , \tag{57}$$

where $I_1(\delta)$ and $A(\delta)$ are the ionization potential and electron affinity of the quark atom. δ represents the extra nucleon charge of the quark attached to the nucleus of charge Z. To estimate $I_1(\delta)$ and $A(\delta)$, Lackner and Zweig interpolated these two quantities within isoelectronic series. These authors have presented an interesting overview of quark chemistry[53].

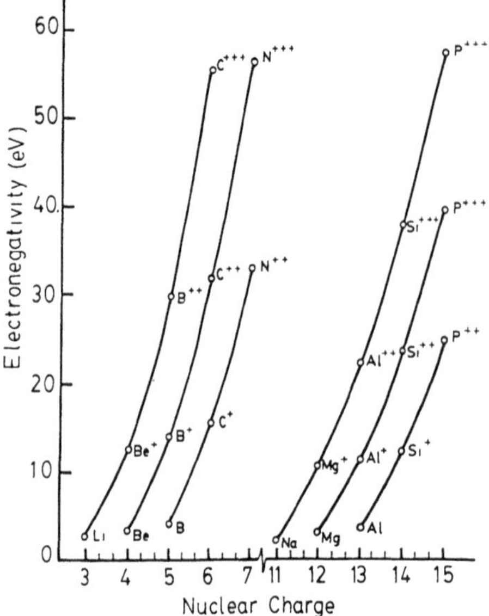

Fig. 7

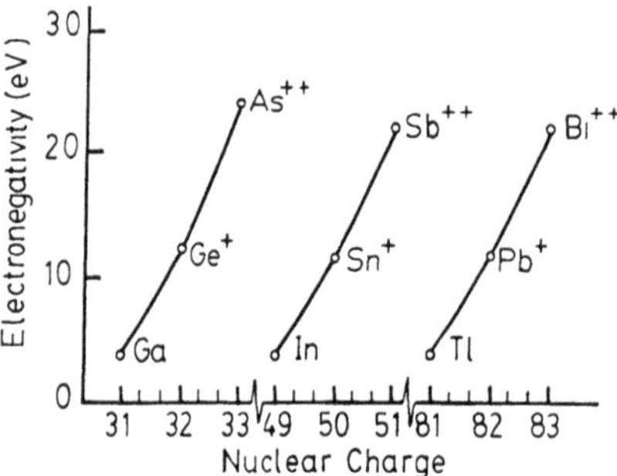

Fig. 8

Figs. 7, 8. Density functional electronegativity of atoms and ions as a function of nuclear charge. Lines join isoelectronic series

Here we calculate the electronegativity of several model quark atoms. In other words, we are not claiming the actual existence of the quark atoms studied in this paper. We are only presenting a method to compute χ in those atoms based on the Weizsäcker-type density functional. Figures 7 and 8 show the density functional electronegativities for several isoelectronic series. The nuclear charge increases across an isoelectronic series. Then the electronegativity of quark atoms or ions can be obtained by quadratic interpolation in each isoelectronic series. Table 3 provides a comparison, for several quark atoms, between the interpolated density functional electronegativity χ_{DF} and the empirical electronegativity χ_{ex}^{M} obtained also by quadratic interpolation using Mulliken electronegativities of atoms and ions calculated from experimental energies[49]. Symboles like (Li)1/3 or (Li)2/3 indicate that particles with positive charge of magnitude 1/3 or 2/3 (in atomic units) are respectively attached to the nucleus of a lithium atom. The electronegativities of quark atoms differ substantially from the electronegativities of the corresponding normal atoms and then have a different chemical behaviour.

Table 3. Density functional electronegativity χ_{DF} and empirical electronegativity χ^{M}_{ex} of several fractionally charged atoms. Both quantities have been obtained by quadratic interpolation in isoelectronic series. Δ is the relative difference $\Delta = 100\,(\chi_{DF} - \chi^{M}_{ex})/\chi^{M}_{ex}$. χ_{DF} and Δ corresponding to normal atoms are given for comparison. Units of χ are eV

	X		(X)1/3			(X)2/3		
Atom (X)	χ_{DF}	Δ	χ^{M}_{ex}	χ_{DF}	Δ	χ^{M}_{ex}	χ_{DF}	Δ
Li	2.42	− 20	5.82	5.07	− 13	9.41	8.49	− 10
Be	3.42	− 25	7.89	6.22	− 21	11.91	9.80	− 18
B	4.13	− 4	8.00	7.27	− 9	12.50	11.07	− 11
Na	2.34	− 18	5.25	4.78	− 9	8.09	7.57	− 6
Mg	3.15	− 14	6.15	5.50	− 11	9.07	8.27	− 9
Al	3.73	16	5.82	6.22	7	8.83	9.13	3
Ga	3.83	20	5.78	6.36	10	8.68	9.22	6
In	3.72	24	5.43	6.10	12	8.08	8.75	8
Tl	3.81	19	5.66	6.31	11	8.34	9.03	8

Table 4. Density functional electronegativity χ_{DF}, empirical electronegativity χ^{M}_{ex} and approximate DF electronegativity χ^{M}_{DF} of several quark ions. Those three quantities have been obtained by quadratic interpolation in isoelectronic series. The relative difference $100\,(\chi_{DF} - \chi^{M}_{ex})/\chi^{M}_{ex}$ and $100\,(\chi^{M}_{DF} - \chi^{M}_{ex})/\chi^{M}_{ex}$ are given in brackets. Units of χ are eV

Ion	χ_{DF}	χ^{M}_{DF}	χ^{M}_{ex}
(Be^{+})1/3	17.56 (− 7)	18.66 (− 1)	18.92
(Be^{+})2/3	23.29 (− 6)	24.51 (− 1)	24.84
(B^{+})1/3	19.30 (− 14)	20.31 (− 9)	22.37
(B^{+})2/3	25.21 (− 12)	26.31 (− 9)	28.80
(C^{+})1/3	20.68 (− 13)		23.30
(C^{+})2/3	26.49 (− 14)		30.81
(Mg^{+})1/3	14.21 (− 6)	14.75 (− 2)	15.04
(Mg^{+})2/3	18.09 (− 5)	18.61 (− 3)	19.13
(Al^{+})1/3	15.10 (− 7)	15.66 (− 3)	16.17
(Al^{+})2/3	19.15 (− 6)	19.71 (− 3)	20.34
(Si^{+})1/3	16.20 (1)		16.06
(Si^{+})2/3	20.35 (0)		20.30
(Ge^{+})1/3	15.98 (3)		15.44
(Ge^{+})2/3	19.87 (3)		19.30
(Sn^{+})1/3	14.82 (6)		14.04
(Sn^{+})2/3	18.24 (5)		17.36
(Pb^{+})1/3	15.07 (5)		14.31
(Pb^{+})2/3	18.41 (5)		17.61
(B^{++})1/3	37.46 (− 4)	39.12 (0)	38.98
(B^{++})2/3	46.00 (− 3)	47.85 (1)	47.19
(C^{++})1/3	39.30 (− 11)	40.63 (− 8)	44.12
(C^{++})2/3	47.48 (− 10)	48.95 (− 7)	52.90
(Al^{++})1/3	27.13 (− 5)	27.69 (− 3)	28.48
(Al^{++})2/3	32.22 (− 4)	32.91 (− 2)	33.71
(Si^{++})1/3	28.52 (− 4)	29.07 (− 3)	29.82
(Si^{++})2/3	33.83 (− 4)	34.39 (− 2)	35.10
(C^{+++})1/3	65.94 (0)	68.36 (4)	65.90
(C^{+++})2/3	77.34 (1)	80.13 (5)	76.40
$Si^{+++})$1/3	44.09 (− 3)	44.73 (− 1)	45.24
(Si^{+++})2/3	50.65 (− 2)	51.34 (0)	51.53

The relative difference $(\chi_{DF} - \chi^M_{ex})/\chi^M_{ex}$ is substantially due to the systematic difference between χ and χ^M, that is, to the fact that χ^M is a finite difference approximation to χ (notice that elements of the N, O and F groups are absent from Table 3). Table 4 provides a similar analysis of the difference between χ and χ^M for quark ions (ions with an extra quark attached to the nucleus). In this table the density functional electronegativity χ_{DF} is compared to the empirical electronegativity χ^M_{ex} and to χ^M_{DF} obtained by the same procedure as χ^M_{ex} but using theoretical energies. Symbols like $(Be^+)1/3$ or $(Be^+)2/3$ indicate that quarks with charge of magnitude 1/3 or 2/3 are respectively attached to the nucleus of a Be^+ ion. Relative differences with respect to χ^M_{DF} are given in brackets. The differences between χ^M_{DF} and χ^M_{ex} are much smaller than the differences between χ_{DF} and χ^M_{ex}. This shows that a substantial part of the difference between χ_{DF} and χ^M_{ex} is due to the finite difference approximation used in evaluating χ^M.

4 Application to Molecules

In this section we illustrate how the use of electronegativity can shed light in understanding molecular binding. A large body of work has been done by many different authors on this topic. Our purpose here is not to review such a vast field, but only to present a few specific examples, mainly taken from our own work. In this context, we enlarge the validity of the applications shown, by sometimes employing, instead of the density functional electronegativities calculated in Sect. 3, empirical electronegativities which evidently lead to better quantitative results.

4.1 Equalization of Electronegativities and Charge Transfer in Molecule Formation

A simple model for the transfer of electronic charge between atoms in a molecule is now considered. We first write the heat of formation of a diatomic XY molecule as

$$\Delta H(XY) = [E(Z_Y, Z_Y + Q) - E(Z_Y, Z_Y)] + [E(Z_X, Z_X - Q) - E(Z_X, Z_X)] - \frac{Q^2}{R} + E_{cov} \,. \qquad (58)$$

where Q is the number of electrons (normally a non-integer number) transferred from X to Y and R is the internuclear separation. That is, if $\chi(Y) > \chi(X)$, then Q is a positive number. The first two terms in (58) give the change in the energy of the atom when their configuration is changed, the third term is the Madelung energy of the molecule and the fourth term is the covalent contribution to the XY bond. Minimization with respect to Q leads to

$$\frac{dE(Z_Y, Z_Y + Q)}{dQ} + \frac{dE(Z_X, Z_X - Q)}{dQ} - \frac{2Q}{R} + \frac{dE_{cov}}{dQ} = 0 \,. \qquad (59)$$

The last two terms of this equation are of opposite sign and tend to cancel each other[80]. That is, an increase in Q will enhance the ionic contribution and will diminish the covalent contribution to the bond. This effect can be taken into account in a simple way by writing

$$\frac{dE(Z_Y, Z_Y + Q)}{dQ} + \frac{dE(Z_X, Z_X - Q)}{dQ} - \frac{2\alpha Q}{R} = 0 . \tag{60}$$

where α is a number small with respect to unity. Since $Q = N_Y - Z_Y = -(N_X - Z_X)$ this equation can be written in a more convenient form for our purposes

$$\left.\frac{\partial E(Z_Y, N_Y)}{\partial N_Y}\right|_{N_Y = Z_Y + Q} + \frac{\alpha(N_X - Z_X)}{R} = \left.\frac{\partial E(Z_X, N_X)}{\partial N_X}\right|_{N_X = Z_X - Q} + \frac{\alpha(N_Y - Z_Y)}{R} , \tag{61}$$

that is

$$\chi(Z_Y, Z_Y + Q) + \frac{\alpha Q}{R} = \chi(Z_X, Z_X - Q) - \frac{\alpha Q}{R} . \tag{62}$$

This proves Sanderson's principle of electronegativity equalization[12, 13, 81] for this simple model. Equation (62) can be interpreted as stating that the electronegativity of an atom in the molecule contains one intraatomic and one interatomic term. Assuming now that $\chi(Z, N)$ varies linearly with N about $N = Z$ (this is equivalent to assuming a parabolic dependence of E with N), then (62) can be written as

$$\chi(Y) + (N_Y - Z_Y)\chi'(Y) + \frac{\alpha Q}{R} = \chi(X) + (N_X - Z_X)\chi'(X) - \frac{\alpha Q}{R} . \tag{63}$$

where $\chi(Y) = \chi(Z_Y, Z_Y)$, $\chi(X) = \chi(Z_X, Z_X)$ and the derivatives $\chi'(X)$ and $\chi'(Y)$ are given by Eq. (44). Equation (63) leads to the following result for the charge transfer (number of electrons transferred from X and Y)

$$Q = -\frac{\chi(Y) - \chi(X)}{\chi'(X) + \chi'(Y) + (2\,\alpha/R)} . \tag{64}$$

Since α is much smaller than unity, the last term in the denominator can be neglected. This is supported by an analysis of the values of Q obtained for several molecules. In this case (64) reduces to

$$Q = -\frac{\chi(Y) - \chi(X)}{\chi'(X) + \chi'(Y)} . \tag{65}$$

which is a result used previously by several workers[7, 23, 50, 82]. Since χ' is negative, then $Q > 0$ if $\chi(Y) > \chi(X)$. On the other hand, the electronegativity $\chi(XY)$ of the molecule (which in this simple model is identified with the electronegativity of the two atomic species after charge transfer) becomes

$$\chi(XY) = \frac{\chi(X)\chi'(Y) + \chi(Y)\chi'(X)}{\chi'(X) + \chi'(Y)} . \tag{66}$$

Equations (64), (65) and (66) also demonstrate the interest in calculating the first derivative of the electronegativitiy, dealt with in Sect. 3 above. The results obtained by keeping also quadratic terms in the expansion of $\chi(Z, N)$ about $N = Z$ have been given in Ref.[23].

If the assumed parabolic dependence of the energy with N is fitted to experimental data, Eqs. (4) and (45) result for χ and χ', as we already know. In that case Eq. (65) becomes

$$Q = -\frac{1}{2}\frac{I_1(X) + A(X) - I_1(Y) - A(Y)}{I_1(X) - A(X) + I_1(Y) - A(Y)} . \tag{67}$$

To judge the validity of Eq. (67) an experimental measure of the charge transfer in the molecule is needed. This is a difficult point since a completely satisfactory way of estimating the charge transfer from experiment does not exist. In other words, the partition of bonding electrons into a fraction belonging to atom X and another fraction belonging to atom Y can not be performed in a unique way. Nevertheless, reasonable measures of charge transfer have been proposed and the concept of charge transfer has been traditionally used in molecular chemistry, where many molecular properties have been related to the degree of electron transfer in the molecule. Due to this difficulty one must be satisfied in correlating trends, rather than absolute values of charge transfer, in classes of related molecules. Well aware of these limitations we have used a crude way of estimating Q from experimentala dat, defined through the relation

$$D = Q_{ex}R , \tag{68}$$

where R and D are the experimental equilibrium internuclear separation and electric dipole moment respectively. Figure 9 gives a comparison of Q_{ex}, obtained from Eq. (68), with the theoretical Q obtained from Eq. (67), for the molecules FH, ClH, BrH, IH, ClBr, BrI, ICl, ClF and BrF. Measured dipole moments, ionization potentials and electon affinities have been taken from Refs. 83, 84 and internuclear distances from Ref. 85. The molecules studied can be separated into two groups: mixed halides and hydrogen halides. It can be seen from Fig. 9 that a satisfactory linear correlation between the theoretical and the empirical charge transfer exists within each group. Even more, the slope of the line for the mixed halides is close to unity. The reason for the two different lines might be linked in part to the estimation of $\chi'(H)$[82].

In Fig. 10 we have plotted $\chi(XY)$ versus $\chi(X) - \chi(Y)$. A set of linear relations is obtained, corresponding to molecules with X fixed and Y variable. This is reminiscent of a similar set of linear relations obtained by Mucci and March[86], who plotted the first ionization potential of the molecule versus the difference in the ionization potentials of the two free atoms. The present results in Fig. 10 would serve as a theoretical explanation of the empirical correlation exposed by Mucci and March if $I_1 \propto \chi$. This is approximately true. In Mulliken's approximation $\chi = (I_1 + A)/2$, and since A is generally small compared to I_1, then $\chi \approx I_1/2$.

Charge transfers in alkaline molecules are plotted in Fig. 11. The results have been obtained from Eq. (65) using density functional values of χ and χ' (Table 2). The calcu-

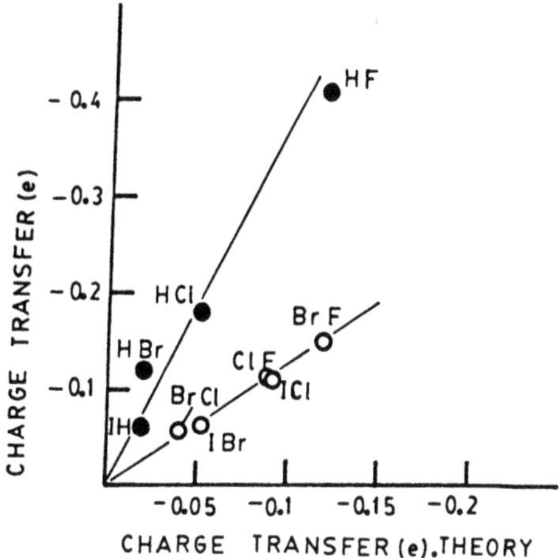

Fig. 9. Empirical charge transfer (Eq. (68)) versus theoretical charge transfer from Eq. (67), in some XY molecules ($Q = N_Y - Z_Y$). Electrons are transferred from atom X to atom Y in these molecules

lated charge transfers are very small in these molecules, in accord with chemical intuition.

The extension of the simple charge transfer theory to polyatomic molecules XY_n with high symmetry is very simple[7)] and the number of electrons lost by atom X is easily found to be (using $\alpha = 0$)

$$Q = -\frac{\chi(Y) - \chi(X)}{\chi'(X) + (\chi'(Y)/n)}\,. \tag{69}$$

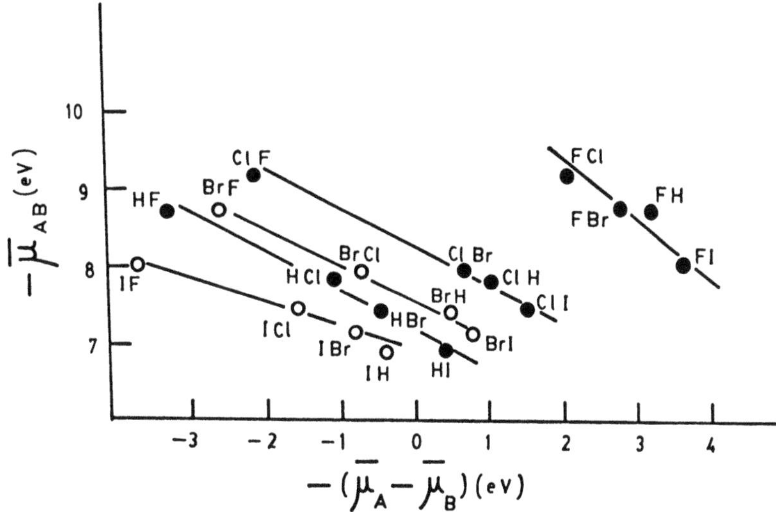

Fig. 10. Electronegativity ($\chi = -\mu$) of the molecule versus the difference of electronegativity between the component atoms

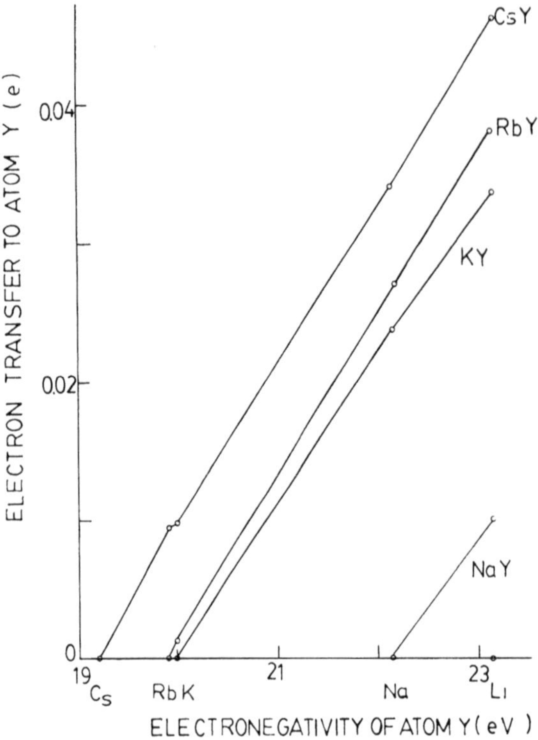

Fig. 11. Calculated charge transfer in CsY, RbY, KY and NaY molecules, with Y = Cs, Rb, K, Na and Li

Q/n electrons are then transferred to each Y atom. On the other hand the electronegativity of the molecule becomes

$$\chi(XY_n) = \frac{\chi'(X)\chi(Y) + \chi(X)(\chi'(Y)/n)}{\chi'(X) + (\chi'(Y)/n)} . \tag{70}$$

These equations have been applied to some tetrahedral XY_4 and octahedral XY_6 molecules. The results are given in Table 5.

Table 5. Electron transfer in tetrahedral XY_4 (X = C, Si, Ge, Sn; Y = F, Cl, Br) and octahedral XY_6 (X = S, Se, Te; Y = F) molecules. Molecular electronegativity (eV) is given in brackets

	F	Cl	Br
C	0.09 (5.395)	0.06 (5.19)	0.05 (5.09)
Si	0.155 (5.24)	0.13 (5.065)	0.11 (4.96)
Ge	0.155 (5.24)	0.13 (5.065)	0.11 (4.96)
Sn	0.175 (5.20)	0.15 (5.03)	0.14 (4.96)
S	0.07 (5.50)		
Se	0.08 (5.48)		
Te	0.10 (5.45)		

4.2 *Simple Charge Transfer Model of X-Ray Scattering by Ten-Electron Molecules*[87]

The differences between the experimental X-ray scattering functions of ten-electron molecules (Ne, OH_2, NH_3, CH_4)[88] reflect systematic changes in the spatial distribution of the electronic cloud. Banyard and March[89] provided a theoretical interpretation based on approximate molecular wavefunctions obtained by the Hartree-Fock method. The approximation consisted in using a spherically symmetric molecular electron density, obtained by solving the HF equations after averaging the nuclear field over angles about the central nucleus. Subsequently, by developing a general method for calculating the X-ray scattering by a gas of non-spherical molecules, Banyard and March[90] showed for OH_2 that corrections (in particular the leading p term) to the spherical average density (s term) had only a very small effect on the scattering factor. Such corrections can therefore be safely neglected for the molecules considered here.

The scattering factor will now be calculated using a model which consists in building the molecule from atomic fragments, each fragment having a total electronic charge different from that of the neutral atom. These charges are calculated using the condition of equal chemical potential of the fragments in the molecule, that is Eq. (69). The interest of the model goes, in our view, beyond the computation of X-ray scattering functions. Since one of the objectives of the theory of the molecular binding is to explain the properties of the molecules in terms of the corresponding properties of the atoms building the molecule, interest exists in decomposing the molecular electron density into localized fragments[91–94].

The electron density $\varrho(\bar{r})$ in a molecule XH_m (in our case OH_2, NH_3 or CH_4) is constructed using the equation

$$\varrho(\bar{r}) = \sum_i \varrho_i(\bar{r} - \bar{R}_i) \,, \tag{71}$$

where $\bar{R}_i$ are the nuclear positions[85] and ϱ_i are the electron densities of the atomic fragments. Using Eq. (69), the number of electrons in the atomic fragments are found to be

$$N(X) = Z(X) - Q \tag{72}$$

$$N(H) = 1 + Q/m \,, \tag{73}$$

where $Z(X)$ is the atomic number of atom X, that is, the number of electrons in the neutral atom. Evaluating χ and χ' from experimental ionization potentials and electron affinities, the following charges Q result, $Q = -0.022$, $Q = 0.011$ and $Q = 0.071$ in OH_2, NH_3 and CH_4 respectively. The electronic configuration of free O, N and C atoms is $1s^2\, 2p^2\, 2p^{6-m}$ and the corresponding free atom electron density, separated in shell contributions, is

$$\varrho_X^0(\bar{r}) = \varrho_{1s}(\bar{r}) + \varrho_{2s}(\bar{r}) + \varrho_{2p}(\bar{r}) \,. \tag{74}$$

The electron density of the fragment X in the molecule is built as

$$\varrho_X(\bar{r}) = \varrho_{1s}(\bar{r}) + \varrho_{2s}(\bar{r}) + \varrho_{2p}^*(\bar{r}) \tag{75}$$

where ϱ_{1s} and ϱ_{2s} are taken the same as in the free atom and the 2p shell density,

$$\varrho^*_{2p}(\bar{r}) = M_X \tilde{\varrho}_{2p}(r) \tag{76}$$

is formed by first averaging $\varrho_{2p}(\bar{r})$ over angles ($\tilde{\varrho}_{2p}(r)$ is the spherical average) and then renormalizing $\tilde{\varrho}_{2p}$ (r) to have the correct number of electrons, that is $6 - m - Q$. In a similar way the electron density of each H fragment is

$$\varrho_H(\bar{r}) = M_H \varrho^0_H(\bar{r}) \; . \tag{77}$$

The normalization constants M_X and M_H are respectively

$$M_X = 1 - \frac{Q}{6 - m} \, , \tag{78}$$

$$M_H = 1 + \frac{Q}{m} \; . \tag{79}$$

The total molecular density is then constructed from Eq. (71), using (75) and (77) and experimental internuclear distances. Analytical HF wave functions[68)] were used to construct ϱ_{1s}, ϱ_{2s} and ϱ_{2p}. Following Banyard and March[89)] a spherically symmetric electron density will be used in the computation of the scattering factors. For this purpose, the

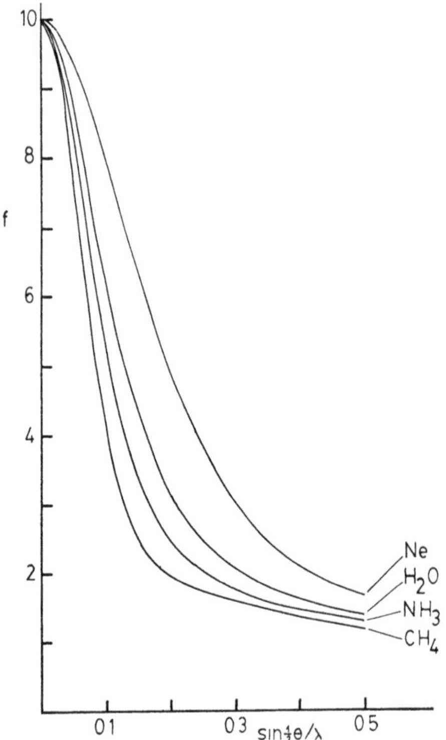

Fig. 12. Calculated scattering factors

molecular electron density $\varrho(\bar{r})$ is spherically averaged over angles about the nucleus of the X fragment. Since $\varrho_X(\bar{r})$ is already spherically symmetric, only the H fragment electron densities need to be averaged. If $\varrho^S(r)$ is the spherically averaged molecular electron density, the scattering factor is obtained as

$$f = \int_0^\infty \varrho^S(r)\, 4\,\pi r^2 \frac{\sin \phi r}{\phi r}\, dr\ , \tag{80}$$

where $\phi = 4\,\pi \sin(\theta/2)/\lambda$; λ is the wavelength of the incident radiation and θ is the scattering angle.

The calculated scattering factors of OH_2, NH_3 and CH_4 are plotted together in Fig. 12, where the systematic differences due to the different spacial distribution of the ten electrons in the molecule are clearly seen. The calculated scattering factor of Neon atom is given for comparison. In Fig. 13 the calculated scattering factor of H_2O is compared to experiment[88] (taken from the figures of Banyard and March[89]) and to the theoretical scattering factor of Banyard and March[89]. A similar comparison for NH_3 and CH_4 is given in Ref. 87. Both theoretical curves provide a good description of the experimental results although the present calculation appears to be slightly more accurate.

Another interesting case in connection with X-ray scattering is the $(NH_4)^+$ ion. As a result of his X-ray analysis of $NH_4\ HF_2$, McDonald[95] has proposed a model for the

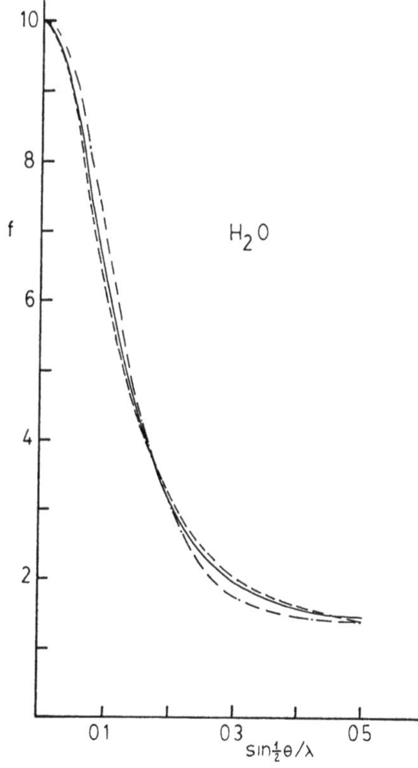

Fig. 13. Scattering factor of H_2O. — Experiment[88, 89]; ----- Theoretical (this work); –·–·– Theoretical results of Banyard and March[89]

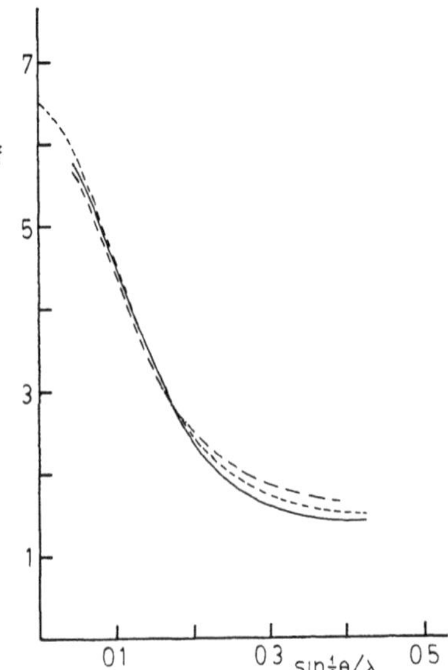

Fig. 14. Scattering factor of the N fragment in $(NH_4)^+$. — Experimental[95]; ----- theoretical (this work, with N(N) = 6.5 e); –·–·– theoretical results of Banyard and March[96] (for $r_0 = 1.4a_0$)

electron distribution in the ammonium ion which can be described as follows. The N nucleus is surrounded by a spherical distribution of charge, but the atom is electron-deficient, containing very nearly six electrons. To this density, McDonald suggests that one should add four hydrogen atoms, with assumed unperturbed density distributions, but rather than centring these on the protons, they should be moved inwards along the N–H bonds, to a distance of 0.86 of the internuclear separation. In the same paper McDonald also gives the measured scattering factor of the N fragment in $(NH_4)^+$ and this result has been plotted in Fig. 14. Banyard and March[96] have performed a theoretical analysis of McDonald's model in the following way: they started from the electron density of $(NH_4)^+$ calculated by Bernal and Massey[97] using the Hartree method. This electron density is spherically symmetric as a consequence of averaging the nuclear field over angles about the N nucleus. Then a spherical density was constructed by placing four unperturbed H atoms at distances r_0 from the N nucleus and averaging the resulting electron density about the center of the molecule. Finally, Banyard and March subtracted the averaged H density from the Bernal-Massey molecular density, obtaining in this way the density corresponding to the N fragment. The result plotted in Fig. 14 corresponds to $r_0 = 1.4\ a_0$ which amounts to 0.76 of the N–H separation.

Since the H cloud motion assumed in McDonald's model can be interpreted as a transfer of charge from H to N^+, we have analyzed the possibility of describing this charge transfer in another way, namely using the same model as for the ten-electron molecules. Application of Eq. (69) to the molecule $(NH_4)^+$, where X is now identified with N^+, gives for the number of electrons of the N fragment in the molecule N(N) = 6.81. The scattering factor of this fragment, calculated using an electron density obtained by renormalizing the free N density as explained above, is in good agreement with the empirical curve of Fig. 14. An even better fit to the empirical curve can be obtained by treating the charge transfer as a parameter. The motivation is that Eq. (69) is not more than an approximation and the same occurs in the calculation of χ (Mulliken approximation) and its first derivative. Thus we have plotted the theoretical scattering factor obtained by using N(N) = 6.5. In the region $\sin(\theta/2)/\lambda > 0.2$ this scattering factor is closer to the empirical one than the scattering factor calculated by Banyard and March.

5 Other Applications

In this section other applications of the electronegativity concept to molecular and solid state physics are presented. Those applications are carried out within the framework of the approximate DF formalism developed in Sect. 2 and already applied to free atoms and ions in Sect. 3.

5.1 Relation of Chemical Potential to Electrostatic Potential and Bonding Distances

Let us again consider the Euler equation of the DF theory

$$\mu = \frac{\delta T[\varrho]}{\delta\varrho(\bar{r})} + \frac{\delta E_{xc}[\varrho]}{\delta\varrho(\bar{r})} + V_T(\bar{r}) . \tag{81}$$

Politzer, Parr and Murphy (PPM)[98] have noticed that at any point $\bar{r}$ at which

$$\frac{\delta T[\varrho]}{\delta\varrho(\bar{r})} = -\frac{\delta E_{xc}[\varrho]}{\delta\varrho(\bar{r})} . \tag{82}$$

then

$$\mu = V_T(\bar{r}) . \tag{83}$$

Equation (83) represents a useful approach to obtain μ in an atom, since the total electrostatic potential $V_T(\bar{r})$ can be obtained from accurate electron densities, provided that one can calculate the points $\bar{r}$ in which Eq. (82) is satisfied. This is a difficult task due to the inaccurate knowledge of the functionals $T[\varrho]$ and $E_{xc}[\varrho]$.

As a starting step, PPM solved Eq. (82) within the TFD theory, which leads to the result that

$$\varrho(\bar{r}_\mu) = 0.00872 \text{ e/a.u.}^3 . \tag{84}$$

$\bar{r}_\mu$ indicates the point where Eq. (82) [or (83)] is fulfilled. But PPM discovered that the use of HF atomic wave functions to evaluate $V_T(\bar{r})$ at points satisfying the universal relation (84) leads to chemical potentials which do not agree with empirical ones. This failure is not surprising and it suggests to solve Eq. (82) for DF theories more accurate than the TFD theory. This is what we do inmediately below. On the other hand, perhaps the most interesting aspect of the work of PPM comes from their observation that the radial distances r_μ at which $V_T(\bar{r})$ (computed from HF atomic wave functions) is equal to the empirical chemical potential

$$V_T^{HF}(r_\mu) = -\frac{I + A}{2}, \tag{85}$$

are rather similar to the empirical single bond covalent radii of the atoms[99]. This suggests that the distance r_μ have a special significance with respect to the bonding properties of atoms.

Since this nice result was obtained by putting together theoretical and empirical information it should be interesting to explore the consequences of Eqs. (82) and (83) fully within an approximate DF theory like the one developed in Sect. 2 (self-consistency is then preserved). In such a case Eq. (82) becomes (see Eq. (35)).

$$\left[\frac{(3\pi^2\varrho)^{2/3}}{2} + \frac{1}{8}\left(\frac{\nabla\varrho}{\varrho}\right)^2 - \frac{1}{4}\frac{\nabla^2\varrho}{\varrho} - \left(\frac{3\varrho}{\pi}\right)^{1/3} - \frac{3.43 + 0.36\,\varrho^{-1/3}}{[7.8 + (4\pi\varrho/3)^{-1/3}]^2}\right]_{r=r_\mu} = 0\,. \tag{86}$$

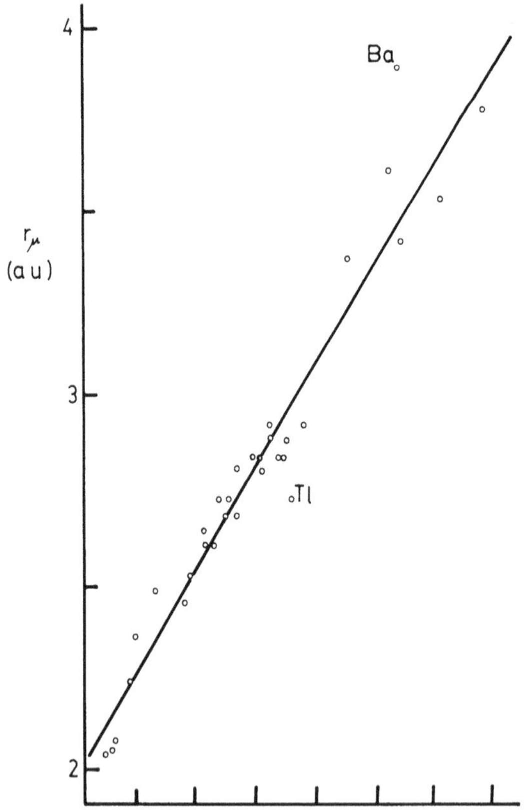

Fig. 15. r_μ versus the empirical non-polar covalent radius r_{cov}[100] for the elements of Table 1

This condition is easily tested after the selfconsistent solution of the Euler equation has been obtained. The calculated values of r_μ have been plotted versus the empirical nonpolar covalent radius r_{cov} for atoms of the groups IA to VIIA, in Fig. 15. The nonpolar covalent radius is defined as half the length of a single bond between uncharged like atoms, and their values have been recently readjusted by Sanderson[100]. A good linear correlation is found

$$r_\mu = 0.55\, r_{cov} + 1.42\ , \tag{87}$$

and only Ba and Tl deviate substantially.

Pressing a little bit further the relation between r_μ and some characteristic bond distance in the condensed phase, we now investigate the relation between r_μ and the theoretical equilibrium (zero pressure) Wigner-Seitz radius R_{WS} (radius of a sphere corresponding to the volume per atom in the solid), calculated within the same approximate DF pseudopotential scheme. Alonso and Iñiguez[101] have obtained a linear relation between R_c and the calculated R_{WS}

$$R_{WS} = 1.44\, R_c + 1.305\ . \tag{88}$$

R_{WS} can then be obtained from the core radii of Table 1. The results have been plotted versus r_μ in Fig. 16 for the groups with valencies equal to 1, 2, 3 and 4. The points are fitted well by a linear relation

$$R_{WS} = 1.692\, r_\mu - 1.961\ . \tag{89}$$

The theoretical values of R_{WS} are, as a trend, smaller than their experimental counterpart. This can be traced back to the simplicity of the pseudopotential used and to our

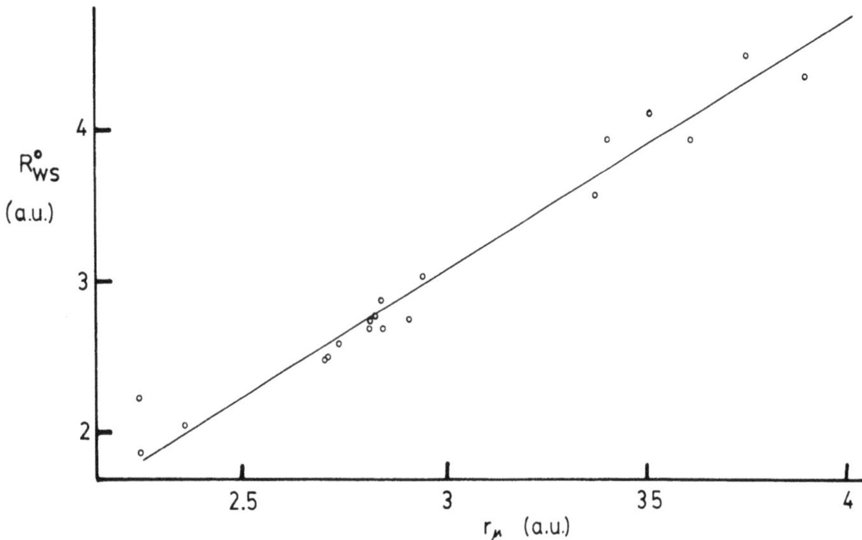

Fig. 16. Calculated equilibrium Wigner-Seitz radius versus r_μ for elements with valencies equal to 1, 2, 3, and 4

particular election of core radii, which have been fitted to a free atom property (see Sect. 3). But the point of interest here is that a linear relation between ab initio r_μ's and R_{WS}'s exists when both quantities are obtained within the same theoretical scheme, pointing again to the significance of r_μ for the characteristic interatomic bonding distance of the atom.

The other point treated by PPM is the justification of Gordy's atomic electronegativity scale[102] Gordy's electronegativity is defined as the electrostatic potential felt by one valence electron at a distance from the nucleus equal to the atom's single bond covalent radius. PPM found a linear relation between μ ($\mu = -(I + A)/2$) and the potential $V_Q(r_\mu)$ due to the charge inside a sphere of radius r_μ

$$V_Q(r_\mu) = -\frac{Q}{r_\mu} . \tag{90}$$

$$Q = Z - \int_{r=0}^{r_\mu} \varrho(\bar{r})\, d\bar{r} . \tag{91}$$

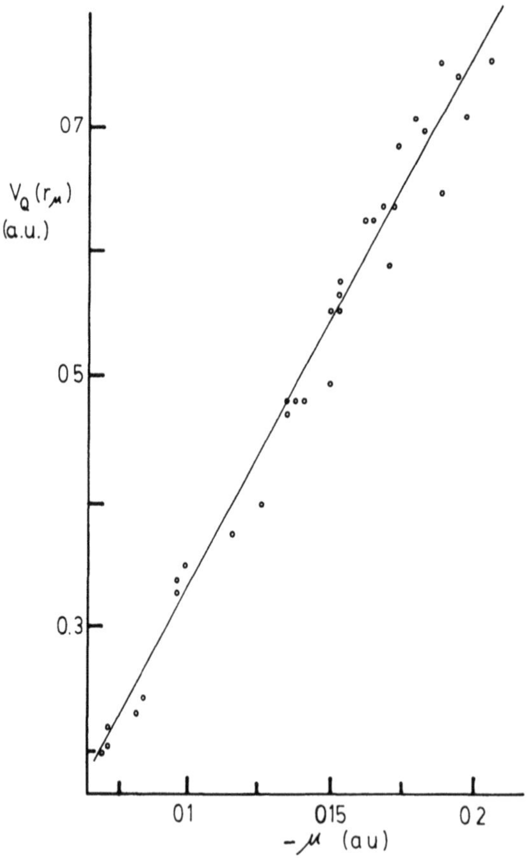

Fig. 17. Electrostatic potential at r_μ due to the change inside a sphere of radius r_μ versus electronegativity, for all the elements in Table 1

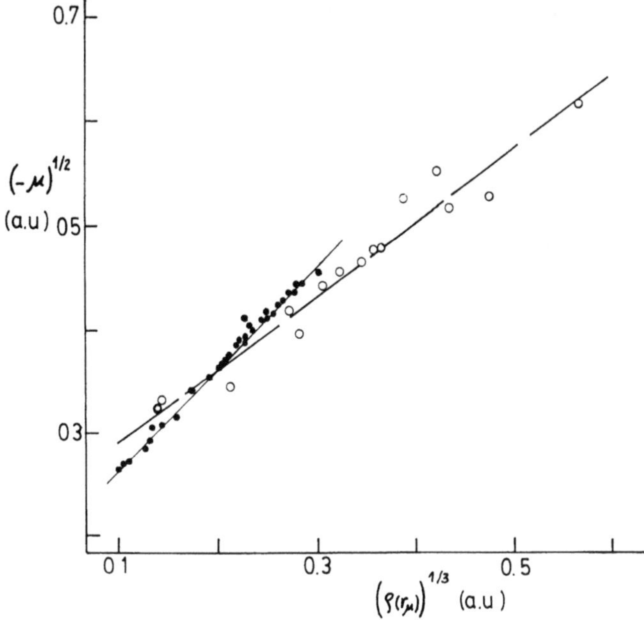

Fig. 18. $(-\mu)^{1/2}$ versus $\varrho(r_\mu)^{1/3}$. Filled circles correspond to the DF calculation. Open circles have been obtained from empirical values of μ and HF values of $\varrho(r_\mu)$

In computing Q, PPM used HF electron densities, and experimental values of I and A were used in evaluating μ. They then found the relation

$$V_Q(r_\mu) = 4.72(-\mu) - 0.118\,. \tag{92}$$

We have repeated this calculation with all the necessary quantities obtained from the approximate DF calculation. The result, plotted in Fig. 17 shows a relation similar to (92), but with slightly different coefficients

$$V_Q(r_\mu) = 4.28(-\mu) - 0.104\,. \tag{93}$$

Finally we have discovered an intriguing empirical correlation between $\varrho(r_\mu)^{1/3}$ and $(-\mu)^{1/2}$, shown in Fig. 18, which includes all the elements of Table 1. A linear fit gives

$$(-\mu)^{1/2} = 0.998\,\varrho(r_\mu)^{1/3} + 0.164\,, \tag{94}$$

with a correlation coefficient equal to 0.990. If the same analysis is performed using Mulliken's values for μ [Eq. (4)] and values of $\varrho(r_\mu)$ obtained from HF electron densities[68] the points (for 15 atoms) can be fitted to the relation

$$(-\mu)^{1/2} = 0.709\,\varrho(r_\mu)^{1/3} + 0.220\,, \tag{95}$$

with a correlation coefficient equal to 0.952.

5.2 *Electronegativity, Ionization Potential and Electron Affinity of Metallic Microparticles*

Small metallic particles are systems of great technological importance, due to their practical applications in catalysis. As a consequence, a growing body of work is presently being dedicated to the study of their electronic and structural properties[103, 104]. An interesting question is the rate at which the value of a given property of a metallic particle converges to its saturation value, that is, the value for a very large particle. This rate depends on the particular property considered. Theoretical calculations can be of great help in this context. The approximate DF formalism developed in previous sections has been applied to this problen[105, 106]. The only difference with respect to Eq. (35) is that the external potential acting on the valence electrons of the microparticle is a sum

$$V_I^P(\bar{r}) = \sum_{\bar{S}} V_I(\bar{r} - \bar{S}) \tag{96}$$

over lattice sites, of ionic pseudopotentials ($V_I(\bar{r} - \bar{S})$ is the pseudopotential of the ion placed at the lattice site $\bar{S}$). The model microparticle is constructed as a small piece of bulk crystal, formed by a central atom surrounded by successive shells of first, second, third etc. neighbors. To simplify the numerical integration of the Euler equation of this problem, the total ionic potential $V_I^P(\bar{r})$ is spherically averaged about the centre of the microparticle. The computation of several electronic and cohesive properties using this model has been presented in detail elsewhere[105, 106]. Here we restrict attention to the chemical potential ($\mu = -\chi$) and to its evolution with the size of the microparticle. Figure 19 compares the evolution of I_1, $-\mu$ and A in sodium microparticles. The points correspond to microparticles with filled coordination shells. Evidently, as $N \rightarrow \infty$ the three quantities go to the same limit (called the work function), but the predicted convergence to a common value is slow. In fact, $(I_1 - A)$ behaves as $1/R_P$, where R_P is the radius of the

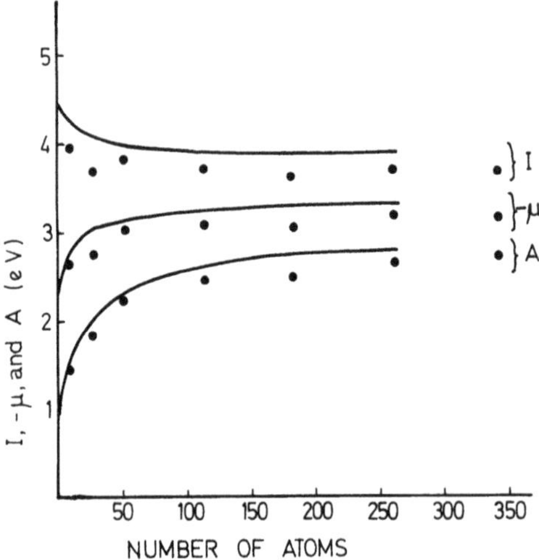

Fig. 19. First ionization potential, electronegativity ($-\mu$) and electron affinity A of sodium microparticles as a function of size. Continuous lines are results of the jellium model. *Filled circles* are results of the pseudopotential model

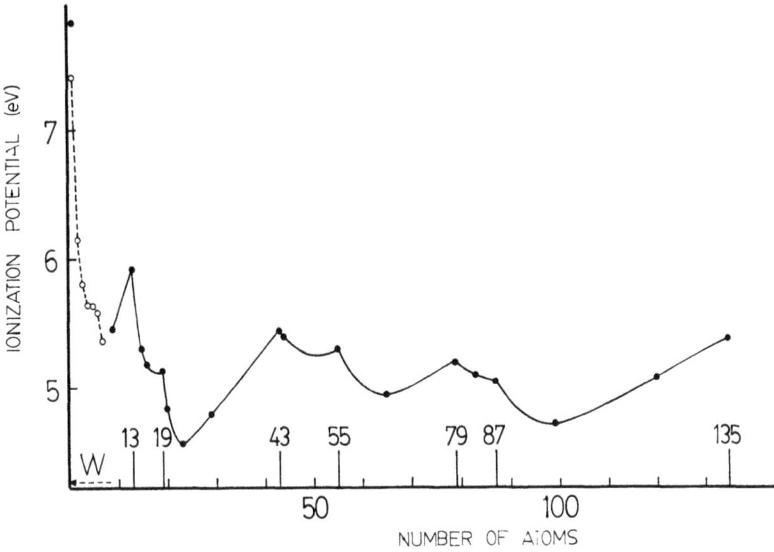

Fig. 20. Ionization potential of lead microparticles as a function of the number of atoms in the particle. Experiment[108] O---O.; this work ●—●. Experimental $I_1(N = \infty)$[109] indicated by W

microparticle[105]. The continuous lines in the figure correspond to a similar calculation, but using a jellium-like model for the ionic background, that is, a positively charged background of constant density, abruptly cut at the surface of the microparticle. I_1, $-\mu$ and A vary smoothly for the jellium-like model particle. This is in contrast to the particle with lattice structure, a fact more clearly seen in Fig. 20, where results for the ionization potential I_1 of lead microparticles are shown[107]. There are oscillations in the value of I_1, with maxima associated with filled coordination shells.

6 Summary

Electronegativity is an important and useful concept in analyzing the chemical binding in molecules and solid compounds. Density functional theory provides a natural route to the study of the electronegativity. The original aim of DF theory is to work with electron density as the basic ingredient. Because of the uncertainties in expressing the energy as a functional of the electron density, effective one-electron wave functions are often introduced into the DF formalism. In this paper, however, we support electron density as the basic ingredient of the theory. We have then applied several simple versions of the DF formalism to study the electronegativities of atoms and ions, showing the virtues and limitations of current approximations to the universal energy density functional. The key of the problem is the treatment of the electronic kinetic energy $T[\varrho]$. A purely local density approximation to $T[\varrho]$ is not enough, and terms taking into account the inhomogeneity of the electron density must be explicitly considered, the form given by Weizsäcker being particularly useful. Functionals based on this approximation are able to give qualitatively (and in some cases semiquantitavely) correct results.

The use of the electronegativity concept has been illustrated by considering several problems related to chemical binding in molecules and solids.

Finally, we mention that the concept of electronegativity, within a DF context, has been extensively used by authors in studies of the chemical binding and cohesive properties of simple metallic alloys. These systems are, however, not considered in this paper and the interested reader should consult the pertinent papers[110–113].

Acknowledgements. This work was supported by CAICYT of Spain (Grant 3265-83).

7 References

1. Pauling, L.: The Nature of the Chemical Bond. Cornell University Press. Ithaca 1960
2. Iczkowski, R. P., Margrave, J. L.: J. Amer. Chem. Soc. *83,* 3547 (1961)
3. Giftopoulos, E. P., Hatsopoulos, G. N.: Proc. Natl. Acad. Sci. USA *60,* 786 (1986)
4. Katriel, J., Parr, R. G., Nyden, M. R.: J. Chem. Phys. *74,* 2397 (1981)
5. Perdew, J. P., Parr, R. G., Levy, M., Balduz, J. L.: Phys. Rev. Lett. *49,* 1691 (1982)
6. Hinze, J., Whitehead, M. A., Jaffe, M. A.: J. Amer. Chem. Soc. *85,* 148 (1962)
7. Ray, N. K., Samuels, L., Parr, R. G.: J. Chem. Phys. *70,* 3680 (1979)
8. Mulliken, R. S.: ibid. *2,* 782 (1934)
9. Komorowski, L.: Chem. Phys. Lett. *103,* 201 (1983)
10. Hohenberg, P., Kohn, W.: Phys. Rev. *136,* B864 (1964)
11. Parr, R. G., Donelly, R. A., Levy, M., Palke, W. E.: J. Chem. Phys. *68,* 3801 (1978)
12. Sanderson, R. T.: Science *114,* 670 (1951)
13. Sanderson, R. T.: Chemical Bonds and Bond Energy. Academic. New York (1971)
14. Thomas, L. H.: Proc. Cambridge Phil. Soc. *23,* 542 (1927)
15. Fermi, E.: Z. Phys. *48,* 73 (1928)
16. March, N. H.: Self-consistent Fields in Atoms. Pergamon. Oxford 1975.
17. Gombás, P.: Handbuch der Physik, Vol. 36 (Ed. Flüge, S.). Springer Verlag. Berlin (1956) p. 109
18. Teller, E.: Rev. Mod. Phys. *34,* 627 (1962)
19. Balazs, N. L.: Phys. Rev. *156,* 42 (1967)
20. Alonso, J. A., Girifalco, L. A.: J. Chem. Phys. *73,* 1313 (1980)
21. Balbás, L. C., Alonso, J. A., Iñiguez, M. P.: Z. Physik *A302,* 307 (1981)
22. Balbás, L. C., Las Heras, E., Alonso, J. A.: ibid. *A305,* 31 (1982)
23. Balbás, L. C., Alonso, J. A., Las Heras, E.: Molec. Phys. *48,* 981 (1983)
24. Balbás, L. C., Alonso, J. A., Del Rio, L. M.: Z. Physik *A312,* 95 (1983)
25. Balbás, L. C., Alonso, J. A., Del Rio, L. M.: Phys. Lett. *101A,* 20 (1984)
26. Balbás, L. C., Vega, L. A., Alonso, J. A.: Z. Phys. *A319,* 275 (1984)
27. Bader, R. F. W., Preston, H. J. T.: Int. J. Quantum Chem. *3,* 327 (1969)
28. Tal, Y., Bader, R. F. W.: Int. J. Quantum Chem. *S12,* 153 (1978)
29. Alonso, J. A., Girifalco, L. A.: Phys. Rev. *B17,* 3735 (1978)
30. Gázquez, J. L., Ludeña, E. V.: Chem. Phys. Lett. *83,* 145 (1981)
31. Gázquez, J. L., Robles, J.: J. Chem. Phys. *76,* 1467 (1982)
    Gázquez, J. L.: Density Functional Theory (Ed. Keller, J., Gázquez, J. L.) Lecture Notes in Physics, Vol. *187,* Springer Verlag. Berlin (1983) p. 229
32. Ludeña, E. V.: J. Chem. Phys. *76,* 3157 (1982)
33. Ludeña, E. V.: Int. J. Quantum Chem. *23,* 127 (1983)
34. Weizsäcker, C. W.: Z. Physik *96,* 431 (1935)
35. Sears, S. B., Parr, R. G., Dinur, U.: Israel J. Chem. *19,* 165 (1980)
36. Acharya, P. K., Bartolotti, L. J., Sears, S. B., Parr, R. G.: Proc. Natl. Acad. Sci. USA *77,* 6978 (1980)

37. Dawson, K. A., March, N. H.: Phys. Lett. *106A,* 158 (1984)
38. Bartolotti, L. J., Acharya, P. K.: J. Chem. Phys. *77,* 4576 (1982)
39. Zorita, M. L., Alonso, J. A., Balbás, L. C.: Int. J. Quantum Chem. *27,* 393 (1985)
40. Deb, B. M., Ghosh, S. K.: ibid. *23,* 1 (1983)
41. Wigner, E. P.: Phys. Rev. *40,* 1002 (1934)
42. Ashcroft, N. W.: Phys. Lett. *23,* 48 (1966)
43. Tomishima, Y., Yonei, K.: J. Phys. Soc. Japan *21,* 142 (1966)
44. Gross, E. K. U., Dreizler, R.: Phys. Rev. *A20,* 1798 (1979)
45. Handy, N. C., Marron, M. T., Silverstone, H. J.: ibid. *180,* 45 (1969)
46. Morrel, M. M., Parr, R. G., Levy, M.: J. Chem. Phys. *62,* 549 (1975)
47. Hoffman-Ostenhof, M., Hoffman-Ostenhof, T.: Phys. Rev. *A16,* 1782 (1977)
48. Tal, Y.: ibid. *A18,* 1781 (1978)
49. Handbook of Chemistry and Physics, 58th ed. C. R. C. Press. Boca Raton 1977
50. Parr, R. G., Pearson, R. G.: J. Amer. Chem. Soc. *105,* 7512 (1983)
51. Pearson, R. G.: ibid. *85,* 3533 (1963)
52. Pearson, R. G.: Science *151,* 171 (1966)
53. Lackner, K. S., Zweig, G.: Phys. Rev. *D28,* 1671 (1983)
54. Robles, J., Bartolotti, L. J.: J. Amer. Chem. Soc. *106,* 3723 (1984)
55. Kohn, W., Sham, L. J.: Phys. Rev. *140,* A1133 (1965)
56. Gunnarsson, O., Jonson, M., Lundqvist, B. I.: ibid. *B20,* 3136 (1979)
57. Bartolotti, L. J., Gadre, S. R., Parr, R. G.: J. Amer. Chem. Soc. *102,* 2945 (1980)
58. Slater, J. C.: The Self Consistent Field for Molecules and Solids. McGraw-Hill, New York 1974
59. March, N. H.: Phys. Lett. *82A,* 73 (1981)
60. March, N. H., Parr, R. G.: Proc. Natl. Acad. Sci. USA *77,* 6285 (1980)
61. Foldy, L. L.: Phys. Rev. *83,* 397 (1951)
62. March, N. H.: Theoretical Chemistry, Vol. 4. Specialist Periodical Reports. Royal Society of Chemistry (1981) p. 92
63. Fraga, S.: Theoret. Chim. Acta *2,* 406 (1964)
64. Politzer, P., Parr, R. G.: J. Chem. Phys. *61,* 4258 (1974)
65. Alonso, J. A., González, D. J., Balbás, L. C.: Int. J. Quantum Chem. *22,* 989 (1982)
66. Balbás, L. C., Zorita, M. L., Alonso, J. A.: ibid. *26,* 145 (1984)
67. Alonso, J. A., March, N. H.: J. Chem. Phys. *78,* 1382 (1983)
68. Clementi, E., Roetti, C.: At. Data Nucl. Data Tables *14,* 177 (1974)
69. La Rue, G. L., Fairbank, W. M., Hebard, W. M.: Phys. Rev. Lett. *38,* 1101 (1977)
70. La Rue, G. S., Fairbank, W. M., Phillips, J. D.: ibid. *42,* 142 (1979)
71. La Rue, G. S., Phillips, J. D., Fairbank W. M.: ibid. *46,* 967 (1981)
72. Hodges, C. L., Abrams, P., Baden, A. R., Bland, R. W., Joyce, D. C., Royer J. P., Walter, F. W., Wilson, E. G., Wong, P. G., Young, K. C.: ibid. *47,* 1651 (1981)
73. Fairbank, W. M., Franklin, A.: Am. J. Phys. *50,* 394 (1982)
74. Lackner, K. S., Zweig, G.: Lett. Nuovo Cimento *33,* 65 (1982)
75. Rank, D. M.: Phys. Rev. *176,* 1635 (1968)
76. Mitsuhasi, Y., Goto, E., Kuroda, R.: J. Phys. Soc. Japan *40,* 613 (1976)
77. Jones, L. W.: Rev. Mod. Phys. *49,* 717 (1977)
78. Gallinaro, G., Marinelli, M., Morpurgo, G.: Phys. Rev. Lett. *38,* 1255 (1977)
79. Marinelli, M., Morpurgo, G.: Phys. Lett. C (Phys. Rep.) *85,* 162 (1982)
80. Hinze, J.: Physical Chemistry, Vol. 5 (Ed. Eyring, H., Henderson, D., Jost, W.) Academic, New York (1970) p. 173
81. Politzer, P., Weinstein, H.: J. Chem. Phys. *71,* 4218 (1979)
82. Alonso, J. A., March, N. H.: Chem. Phys. *76,* 121 (1983)
83. Handbook of Chemistry and Physics, 52nd ed. The Chemical Rubber Company. Cleveland 1971–1972
84. Nethercot, A. H.: Chem. Phys. Lett. *39,* 346 (1978)
85. Tables of interatomic distances and configurations in molecules and ions (Ed. Mitchell, A. D., Cross, L. C.) Special Publication No 11. The Chemical Society. London 1958
86. Mucci, J. F., March, N. H.: J. Chem. Phys. *75,* 5789 (1981)
87. Alonso, J. A., Balbás, L. C., March, N. H.: Molec. Phys. *50,* 789 (1983)

88. Thomer, G.: Phys. Z. *38,* 48 (1937)
89. Banyard, K. E., March, N. H.: Acta Crystallogr. *9,* 385 (1956)
90. Banyard, K. E., March, N. H.: J. Chem. Phys. *26,* 1416 (1957)
91. March, N. H.: Localization and Delocalization in Quantum Chemistry, Vol. I (Ed. Chalvet, O.) Reidel, Dordrecht (1975) p. 115
92. Alonso, J. A., March, N. H.: J. Chem. Phys. *79,* 1903 (1983)
93. Bader, R. F. W., Nguyen-Dang, T. T.: Adv. Quantum Chem. *14,* 63 (1981)
94. Bader, R. F. W., Nguyen-Dang, T. T., Tal, Y.: Rep. Progr. Phys. *44,* 893 (1981)
95. McDonald, T. R. R.: Acta Crystallogr. *13,* 113 (1960)
96. Banyard, K. E., March, N. H.: ibid. *14,* 357 (1961)
97. Bernal, M. J. M., Massey, H. S. W.: Month. Not. Roy. Astr. Soc. *114,* 172 (1954)
98. Politzer, P., Parr, R. G., Murphy, D. R.: J. Chem. Phys. *79,* 3859 (1983)
99. Huheey, J. E.: Inorganic Chemistry. Harper and Row, New York 1972
100. Sanderson, R. T.: J. Amer. Chem. Soc. *105,* 2259 (1983)
101. Alonso, J. A., Iñiguez, M. P.: Physica *103B,* 333 (1981)
102. Gordy, W.: Phys. Rev. *69,* 604 (1946)
103. Perenboom, J. A. A. J., Wyder, P., Meier, F.: Phys. Lett. C (Phys. Rep.) *78,* 173 (1981)
104. Proc. 2nd Internat. Meeting on the Small Particles and Inorganic Clusters (Ed. Borel, J. P., Buttet, J.) Surf. Sci. *106,* 1–608 (1981)
105. Iñiguez, M. P., Baladrón, C., Alonso, J. A.: ibid. *127,* 367 (1983)
106. Baladrón, C., Iñiguez, M. P., Alonso, J. A.: Solid. St. Commun. *50,* 549 (1984)
107. Baladrón, C., Iñiguez, M. P., Alonso, J. A.: Z. Physik B59, 187 (1985)
108. Saito, Y., Yamauchi, K., Mihama, K., Noda, T.: Japan. J. Appl. Phys. *21,* L396 (1982)
109. Michaelson, H. B.: IBM J. Res. Develop. *22,* 72 (1978)
110. Alonso, J. A., Girifalco, L. A.: Phys. Rev. B19, 3889 (1979)
111. Alonso, J. A., González, D. J., Iñiguez, M. P.: Solid St. Commun. *31,* 9 (1979)
112. Iñiguez, M. P., Alonso, J. A., González D. J., Zorita, M. L.: J. Phys. F: Metal Phys. *12,* 1907 (1982)
113. González D. J., Alonso, J. A.: J. de Physique *44,* 229 (1983)

Fukui Function, Electronegativity and Hardness in the Kohn-Sham Theory

José L. Gázquez, Alberto Vela and Marcelo Galván

Departamento de Química, Divisíón de Ciencias Básicas e Ingeniería, Universidad Autónoma Metropolitana-Iztapalapa, C.P. 55-534, México 13, D.F., 09340, México

This work represents a unified approach to the description of important properties such as electronegativity, hardness and Fukui function in atoms.

The extension to molecular systems may prove very useful to understand some aspects of chemical reactivity.

I. Introduction 80

II. Kohn-Sham Theory 81

III. Fukui Function and Spin-Density 88

IV. Chemical Reactivity 95

V. Conclusions 96

VI. References 97

I. Introduction

There is no doubt that the recent progress in density functional theory has provided a very useful tool for understanding molecular properties and for describing the behaviour of atoms in molecules[1-25]. Concepts such as electronegativity and hardness, which have long been part of the vocabulary of the chemist to describe many aspects of molecular structure, but which were rather intuitive in nature, have been shown to be fundamental and very well defined in a thermodynamic-like description of chemical phenomena. On the other hand, new concepts such as the frontier or Fukui functions, may lead to a powerful and unified criterion for chemical reactivity.

Thus, it has been possible to establish[1,2] that electronegativity[27] (χ), which is the negative of the slope of the ground-state electronic energy (of an atom or a molecule) as a function of the number of electrons, is the negative of the chemical potential (μ) of the density functional theory of Hohenberg and Kohn, that is

$$\chi = -\mu = -\left(\frac{\partial E}{\partial N}\right)_v \tag{1}$$

where v is the external potential. Through this identification, it has been possible to calculate systematically within the same context, atomic ionization potentials, electronegativities and electron affinities[6,12,13]. In particular, Bartolotti, Gadre and Parr[6] used the transition-state concept within X_α theory to obtain an expression for the electronegativity in terms of the X_α Lagrange multipliers. Their results were in good agreement with other electronegativity scales.

On the other hand, Parr and Pearson[10] have defined the absolute hardness of an atom or a molecule as

$$\eta = \frac{1}{2}\left(\frac{\partial^2 E}{\partial N^2}\right)_v \tag{2}$$

and through the use of Eqs. (1) and (2) it has been possible to derive theoretically the principle of hard and soft acids and bases, and to find relations that quantify the amount of charge transferred in the formation of a molecule[10,23,24].

More recently, the definition of the frontier or Fukui function as the derivative of the charge density with respect to the number of electrons[25],

$$f(r) = \left(\frac{\partial \varrho(r)}{\partial N}\right)_v \tag{3}$$

has allowed to prove the frontier orbital theory[26], by assuming that chemical reactions are preferred from the direction which produces maximum initial chemical potential response of a reactant. This in turn implies that the local values of $f(r)$ provide a measure of reactivity. The Fukui function is different for electrophilic, nucleophilic or free radical attack, and reduces to the classical frontier orbitals when frozen core approximations are imposed[28].

The properties described through Eqs. (1)–(3) enter into the the relations[1)]

$$dE = \mu dN + \int \varrho(r) dv(r) dr \tag{4}$$

and[25)]

$$d\mu = 2\eta dN + \int f(r) dv(r) dr \tag{5}$$

(dv(r) represents the change in the external potential) which are fundamental for the description of a time-independent chemical event. Thus, the concepts of electronegativity, hardness and frontier orbitals, which have been extensively used as reactivity criteria, are placed in a very solid theoretical framework.

Since all the relations mentioned above come from density functional theory, it seems appropriate to study them from this point of view. The object of this chapter is to present some results that have been obtained in this direction, in relation with the Kohn-Sham formalism[29, 30)].

The structure of this work is as follows. In Sect. II we shall develop the expressions corresponding to Eqs. (1)–(3) for the Kohn-Sham energy functional, relaxation effects will be discussed mainly in connection with the hardness and the Fukui function. Next, in Sect. III we will show that the Fukui function is directly related to the spin-density, and finally in Sect. IV we will consider some aspects of chemical reactivity in terms of the spin-density.

II. Kohn-Sham Theory

The energy functional in the Kohn-Sham method, for the spin-nonpolarized case is given by[30)]

$$\begin{aligned} E[n_k, \phi_k] = {} & \sum_i n_i \int \phi_i^*(r) \left[-\frac{1}{2} \nabla^2 \right] \phi_i(r) dr \\ & - Z \int \frac{\varrho(r)}{r} dr + \frac{1}{2} \int\int \frac{\varrho(r_1)\varrho(r_2)}{r_{12}} dr_1 dr_2 + \int \varrho(r) \varepsilon_{xc}[\varrho(r)] dr \end{aligned} \tag{6}$$

where n_i is the occupation number of the orbital ϕ_i, $\varepsilon_{xc}[\varrho(r)]$ is the exchange-correlation energy functional, and

$$\varrho(r) = \sum_i n_i \phi_i^*(r) \phi_i(r) \tag{7}$$

This functional avoids the problem of the kinetic-energy functional[31)] by making use of the independent particle approximation. This means that the exchange-correlation energy functional in Eq. (6) contains some part of the kinetic energy, namely

$$\begin{aligned} V_{xc}[\varrho(r)] &= \int \varrho(r) v_{xc}[\varrho(r)] dr \\ &= E_{xc}[\varrho(r)] + T[\varrho(r)] - \sum_i n_i \int \phi_i^*(r) \left[-\frac{1}{2} \nabla^2 \right] \phi_i(r) dr \end{aligned} \tag{8}$$

where T and E_{xc} are the exact kinetic and exchange correlation energy functionals. Thus, to apply the Kohn-Sham formalism, one only needs to introduce one of the several approximations to ε_{xc} that already exist in the literature[40]. Among these, the one proposed by Gunnarsson and Lundqvist[32)] provides a reasonably good description of the electronic structure of atoms, molecules and solids, and so it will be the one used throughout all the calculations presented in this work. However, all the relations can be formally developed without having to specify the explicit form of ε_{xc}.

Now, the minimization of the total energy with respect to the orbitals, subject to the orthonormality constraints, leads to one-electron equations of the form

$$\left\{-\frac{1}{2}\nabla^2 - \frac{Z}{r} + \int \frac{\varrho(r_2)}{|r - r_2|}\, dr_2 + v_{xc}[\varrho(r)]\right\}\phi_i(r) = \varepsilon_i \phi_i(r) \tag{9}$$

where

$$v_{xc}[\varrho(r)] = \frac{\delta(\varrho(r)\varepsilon_{xc}[\varrho(r)])}{\delta\varrho(r)} \tag{10}$$

This set of one-electron equations has to be solved self-consistently to determine ε_i and ϕ_i, which allows us to determine the charge density through Eq. (7).

Using Eqs. (6), (7) and (9) one may determine electronegativities, hardnesses and Fukui functions by recognizing that a change in the number of electrons corresponds, in the Kohn-Sham formalism, to a change in the occupation of the highest atomic orbital. However, it should be noted that although this theory interpolates smoothly the total energy between integer values of N, the real curve for isolated atoms at 0 °K has slope discontinuities[16)]. Therefore, for the evaluation of the derivatives at the neutral atom, one should consider the process $A^{\delta+} \rightarrow A^{\delta-}$ in the limit $\delta \rightarrow 0$, which requires E to be smooth only about $N + \delta$ and $N - \delta$, and dictates which are the relevant orbitals according to the configuration of the neutral system[2,12)]. Thus, for some cases there may be more than one orbital involved in the process.

Taking into account all the considerations mentioned above, it is found that the easiest case corresponds to those open shell atoms in which a single orbital is involved, as shown in the following diagram:

i		X	XX
	XX	XX	XX
	positive ion	neutral atom	negative ion

In the spin non-polarized case this represents the most common situation, and here we will restrict ourselves to such atoms.

Now, Janak[33)] has shown that the variation of the total enery, as given by Eq. (6), with respect to one of the orbital occupations, allowing the orbitals to relax, leads to

$$\left(\frac{\partial E}{\partial n_i}\right) = \varepsilon_i \tag{11}$$

if ε_i is obtained from Eq. (9). Thus, making use of this result one may calculate the electronegativity and the hardness of an atom, including relaxation effects, through the following relations:

$$\chi = -\mu = -\left(\frac{\partial E}{\partial N}\right)_v = -\varepsilon_H \tag{12}$$

and[34)]

$$\eta = \frac{1}{2}\left(\frac{\partial^2 E}{\partial N^2}\right)_v = \frac{1}{2}\iint \frac{\phi_H^*(r_1)\phi_H(r_1)f(r_2)}{r_{12}}\,dr_1dr_2 + \frac{1}{2}\int \phi_H^*(r)\phi_H(r)\,\frac{\delta v_{xc}[\varrho(r)]}{\delta\varrho(r)}\,f(r)dr \tag{13}$$

(the subscript H stands for the highest occupied atomic orbital). The Fukui function can be calculated from Eq. (7) and is given by[28)]

$$f(r) = \left(\frac{\partial\varrho(r)}{\partial N}\right)_v = \phi_H^*(r)\phi_H(r) + \sum_i n_i \frac{\partial[\phi_i^*(r)\phi_i(r)]}{\partial N} \tag{14}$$

Although these three equations are exact within the Kohn-Sham method, they require information which, except for the electronegativity, is not directly given by a neutral atom calculation. Thus, in practice a simpler approach to obtain the values of the hardness and of the Fukui function would require that we calculate ε_H and $\varrho(r)$ for several occupations of the highest occupied atomic orbital which describes the process $A^{\delta+} \rightarrow A^{\delta-}$, to calculate the derivative of each one, with respect to N numerically. Since such approach implies a great computational effort, one could also consider approximations to relate η and $f(r)$ with other properties which can be obtained directly from the neutral atom calculation.

The most direct approximation is to neglect the relaxation effects. Thus, under these conditions Eq. (14) becomes

$$f(r) \cong \phi_H^*(r)\phi_H(r) \tag{15}$$

which identifies the Fukui function with the density of the highest occupied atomic orbital[25)]. From the point of view of the local values, one could expect such approximation to be well away from the nucleus, in the region dominated by the valence orbitals. However, it should be a poor approximation close to the nucleus where the core orbitals dominate.

Substituting Eq. (15) in Eq. (13), one obtains that[13)]

$$\eta \cong \frac{1}{2}\iint \frac{\phi_H^*(r_1)\phi_H(r_1)\phi_H^*(r_2)\phi_H(r_2)}{r_{12}}\,dr_1dr_2 + \frac{1}{2}\int \phi_H^*(r)\phi_H(r)\frac{\delta v_{xc}[\varrho(r)]}{\delta\varrho(r)}\,\phi_H^*(r)\phi_H(r)dr = \frac{1}{2}\,[J(H) + K(H)]\,. \tag{16}$$

and this way, the approximate evaluation of $f(r)$ and η would only require information directly related to the neutral atom calculation.

However, the results provided by Eqs. (15) and (16) would be very poor for those systems in which relaxation effects are important. Therefore, one should try to look for other relationships. Recently Gázquez and Ortiz[13)] developed a simple model based on screening effects that leads to

$$\eta \cong \frac{1}{4} \int \frac{\phi_H^*(r)\phi_H(r)}{r}\, dr = \frac{1}{4} \langle r^{-1} \rangle_H \tag{17}$$

This relation leads to values of the hardness that lie very close to those obtained from Eq. (13). This implies that Eq. (17) partially includes relaxation effects. In fact, there is an interesting explanation in relation with this equation. The argument starts by considering the chemical potential as a function of Z and N, which allows us to write

$$d\mu = \left(\frac{\partial\mu}{\partial Z}\right)_N dZ + \left(\frac{\partial\mu}{\partial N}\right)_Z dN \tag{18}$$

Now consider the variation of μ along the neutral atom path, then from Eq. (18) we have that

$$\left(\frac{\partial\mu}{\partial Z}\right)_{N=Z} = \left(\frac{\partial\mu}{\partial Z}\right)_N + \left(\frac{\partial\mu}{\partial N}\right)_Z \tag{19}$$

For those models of atomic structure which predict $\mu = 0$ for the neutral system (like the Thomas-Fermi theory), the derivative on the left-hand side of Eq. (19) would be zero, and therefore

$$\left(\frac{\partial\mu}{\partial N}\right)_Z \cong -\left(\frac{\partial\mu}{\partial Z}\right)_N \tag{20}$$

Using Eqs. (2)–(3), and the Hellmann-Feynman theorem one is led to

$$\eta \cong \frac{1}{2} \int \frac{f(r)}{r}\, dr = \frac{1}{2} \langle r^{-1} \rangle_f \tag{21}$$

which links the hardness with the electrostatic potential at the nucleus of the Fukui function[23)]. We have found[35)], using a finite-differences approach[24)] and Hartree-Fock values to calculate the derivatives of Eq. (20), that Eq. (21)) is approximately valid for all open shell atoms in the periodic table, up to $Z = 85$.

The relation between Eqs. (17) and (21) can be easily established by making use of Eq. (15). Since both equations lead to similar values of the hardness, the difference in the numerical factor (1/4 in the case of Eq. (17) and 1/2 in the case of Eq. (21)) implies that

$$\int \frac{f(r)}{r}\, dr \cong \frac{1}{2} \int \frac{\phi_H^*(r)\phi_H(r)}{r}\, dr \tag{22}$$

Table 1. Comparison of the quantities involved in Eq. (22) for several atoms (atomic units)[a]

Atom	$\langle r^{-1}\rangle_f$	$\langle r^{-1}\rangle_H$	Ratio[b]
Li	0.269	0.358	0.75
B	0.361	0.604	0.60
C	0.434	0.775	0.56
N	0.503	0.944	0.53
O	0.573	1.110	0.52
F	0.641	1.275	0.50
Na	0.245	0.327	0.75
Al	0.257	0.396	0.65
Si	0.302	0.490	0.62
P	0.346	0.579	0.60
S	0.388	0.665	0.58
Cl	0.429	0.750	0.57
K	0.199	0.265	0.75
Br	0.381	0.631	0.60
Rb	0.186	0.247	0.75
I	0.327	0.522	0.63

[a] All integrals are defined in the text; [b] Ratio = $\langle r^{-1}\rangle_f/\langle r^{-1}\rangle_H$

In order to test this and other relations quoted in this work, we have done calculations of several open shell atoms using the Kohn-Sham method with the exchange-correlation approximation of Gunnarsson and Lundqvist[32)]. The exact values of $f(r)$ and η were obtained by calculating each atom for $Z - N = 0.2, 0.1, 0.0, -0.1$ and -0.2 and then performing numerical differentiation.

All other quantities can be obtained directly from the information provided by the neutral atom calculation.

Table 2. Comparison between different approximations to the hardness of several atoms (eV)

Atom	Eq. (13)	Eq. (17)	R_1[a]	Eq. (21)	R_2[b]
Li	2.36	2.43	0.97	3.67	0.64
B	4.08	4.11	0.99	4.92	0.83
C	5.00	5.27	0.95	5.91	0.85
N	5.90	6.42	0.92	6.84	0.86
O	6.77	7.55	0.90	7.80	0.87
F	7.62	8.67	0.88	8.72	0.87
Na	2.30	2.22	1.04	3.33	0.69
Al	2.80	2.69	1.04	3.50	0.80
Si	3.40	3.33	1.02	4.11	0.83
P	3.96	3.94	1.01	4.71	0.84
S	4.49	4.53	0.99	5.27	0.85
Cl	5.00	5.10	0.98	5.83	0.86
K	1.92	1.80	1.07	2.71	0.71
Br	4.48	4.29	1.04	5.18	0.86
Rb	1.83	1.68	1.09	2.53	0.72
I	3.87	3.55	1.09	4.45	0.87

[a] R_1 is the ratio between the values obtained through Eqs. (13) and (17)
[b] R_2 is the ratio between the values obtained through Eqs. (13) and (21)

Thus, in Table 1, one can see the approximate validity of Eq. (22). In general the ratio is greater than the predicted value of 1/2, but the general trend of a constant ratio is reasonably obeyed.

Now, one can also compare the values of η obtained through the different approximations described here. This is done in Table 2, where it may be seen that Eqs. (17) and (21) provide reasonable values of the hardness.

So far we have only made comparisons among different approximations, all of them within the Kohn-Sham method. Thus, in order to compare these results with experimental information, one may use the values of χ and η to calculate the first ionization potential and the electron-affinity. This can be done through the finite-difference formulas for the electronegativity and hardness, namely

$$\chi = \frac{I + A}{2} \tag{23}$$

and

$$\eta = \frac{I - A}{2} \tag{24}$$

which may be combined to obtain

$$I = \chi + \eta \tag{25}$$

and

$$A = \chi - \eta \tag{26}$$

The results obtained through these equations are reported in Table 3. It may be seen that there is a very good agreement between the first ionization potential calculated by the transition state method[30)], and the one calculated from Eq. (25) with χ given by Eq. (12) and η given by Eq. (13). The results for the electron affinity are less accurate with respect to the experimental values, but slightly better than those of Bartolotti, Gadre and Parr[6)]. Nevertheless, the values are good to about 1 eV, and the trends predicted are quite reasonable.

Now, it is interesting to note that since $I - A = 2\eta$, then using the several approximations to η, one obtains the following relations:

From Eq. (13),

$$I - A \cong \iint \frac{\phi_H^*(r_1)\phi_H(r_1)f(r_2)}{r_{12}}\, dr_1 dr_2 + \int \phi_H^*(r)\phi_H(r)\, \frac{\delta v_{xc}[\varrho(r)]}{\delta\varrho(r)}\, f(r)dr \tag{27}$$

From Eq. (16) neglecting the second integral, which is smaller than the first one,

$$I - A \cong \iint \frac{\phi_H^*(r_1)\phi_H(r_1)\phi_H^*(r_2)\phi_H(r_2)}{r_{12}}\, dr_1 dr_2 \tag{28}$$

Table 3. First ionization potentials (I) and electron affinities (A) for several atoms (eV)[a]

Atom	I			A		
	TS[b]	Eq. (25)[c]	Exp	$2\chi - I$[d]	Eq. (26)[c]	Exp
Li	5.03	5.35	5.39	0.12	0.63	0.62
B	7.92	8.00	8.30	−1.12	−0.16	0.28
C	10.63	10.66	11.26	−0.37	0.66	1.27
N	13.40	13.42	14.54	0.54	1.62	0.00
O	16.24	16.27	13.61	1.60	2.73	1.46
F	19.19	19.25	17.42	2.81	4.01	3.40
Na	4.70	5.23	5.14	−0.05	0.63	0.55
Al	5.22	5.75	5.98	−0.72	0.15	0.44
Si	7.18	7.77	8.15	0.01	1.66	1.39
P	9.16	9.79	10.48	0.86	1.87	0.75
S	11.21	11.86	10.36	1.82	2.88	2.08
Cl	13.32	13.99	13.01	2.90	3.99	3.62
K	3.89	4.44	4.34	−0.05	0.60	0.50
Br	11.86	12.78	11.84	2.62	3.82	3.36
Rb	3.65	4.26	4.18	−0.08	0.60	0.49
I	10.42	11.42	10.45	2.48	3.68	3.06

[a] Experimental values taken from Ref. 19
[b] Values calculated through the transition-state appropriate for the first ionization
[c] Using the "exact" values of χ and η as given by the Kohn-Sham method
[d] Values taken from Ref. 6. χ and I are calculated separately by the X_α method using the appropriate transition-state in each case

From Eq. (17),

$$I - A \cong \frac{1}{2} \int \frac{\phi_H^*(r)\phi_H(r)}{r}\, dr \tag{29}$$

Table 4. Comparison between the experimental and calculated values of I − A (eV)

Atom	Eq. (27)	Eq. (28)	Eq. (29)	Eq. (30)	Exp[a]
Li	4.72	6.35	4.86	7.34	4.77
B	8.16	11.16	8.22	9.84	8.02
C	10.00	14.24	10.54	11.82	9.99
N	11.80	17.27	12.84	13.68	14.54
O	13.54	20.26	15.10	15.60	12.15
F	15.24	23.22	17.34	17.44	14.02
Na	4.60	6.11	4.44	6.66	4.59
Al	5.60	7.30	5.38	7.00	5.54
Si	6.80	8.94	6.66	8.22	6.76
P	7.92	10.50	7.88	9.42	9.73
S	8.98	12.00	9.06	10.54	8.28
Cl	10.00	13.47	10.20	11.66	9.39
K	3.84	5.07	3.60	5.42	3.84
Br	8.96	11.75	8.58	10.36	8.48
Rb	3.66	4.79	3.36	5.06	3.69
I	7.74	9.93	7.10	8.90	7.39

[a] See footnote a of Table 3

And from Eq. (21),

$$I - A \cong \int \frac{f(r)}{r} dr \tag{30}$$

One may note that Eq. (28) is just the Pariser formula[36, 37] for self-repulsion, which is so important in semiempirical theories of the electronic structure of molecules. Thus Eq. (27), (29) and (30) provide an alternative for evaluating the important quantity I − A. In Table 4 we present the results obtained through these relations. It may be seen that in general they provide a reasonably good representation of I − A, and may therefore be used as an alternative for evaluating some of the parameters used in semiempirical theories[38]. In particular, Eqs. (27) and (30) which are expressed in terms of the Fukui function, already contain information concerning the response of the system to a change in the occupation of the highest orbital, and they can be of great value in semiempirical theories. For example, the iterative Extended-Hückel method[39] requires a knowledge of the valence-orbital ionization energies as a function of charge. Such information could be directly obtained from the relations derived in this work.

In general one can see that the values of the electronegativity, the hardness and the Fukui function, which are very important in the description of atoms in molecules, and in the description of chemical bond and reactivity, can be evaluated in a very simple way within the Kohn-Sham formalism. The values obtained from this theory seem to be in reasonable agreement with other empirical scales, but they have the advantage that they are all determined within the same context; this provides a unified approach to them.

III. Fukui Function and Spin-Density

So far we have only talked about global quantities (integrals over the whole space). However the Fukui function as defined by Eq. (3) contains local information which is intimately linked to chemical reactivity[25]. Thus, in this section we wish to analyse from a local point of view, the Fukui function of open shell atoms. Before doing so we will first establish a relation between $f(r)$ and the spin-density.

All the equations concerned with the Kohn-Sham method described in the previous section refer to the spin non-polarized case. However, it is well known[30] that one can improve the description of the electronic structure of atoms, molecules and solids, by breaking the charge density into spin components. That is,

$$\varrho(r) = \varrho_{\uparrow}(r) + \varrho_{\downarrow}(r) \tag{31}$$

where $\uparrow$ refers to spin up or α, and $\downarrow$ to spin down or β.

Within density functional theory, such an approach indicates that the total energy will be a functional of $\varrho_{\uparrow}$ and $\varrho_{\downarrow}$ instead of the total charge density. Thus, the Kohn-Sham functional becomes

$$\begin{aligned} E[n_{\sigma k}, \phi_{\sigma k}] = & \sum_{\sigma} \sum_{i} n_{\sigma i} \int \phi^{*}_{\sigma i}(r) \left[-\frac{1}{2} \nabla^2 \right] \phi_{\sigma i}(r) dr \\ & - Z \int \frac{\varrho(r)}{r} dr + \frac{1}{2} \int\int \frac{\varrho(r_1)\varrho(r_2)}{r_{12}} dr_1 dr_2 + \sum_{\sigma} \int \varrho_{\sigma}(r) \varepsilon^{\sigma}_{xc}[\varrho_{\sigma}(r)] dr \end{aligned} \tag{32}$$

where $\sigma = \uparrow, \downarrow$, $n_{\sigma i}$ is the occupation of the spin-orbital $\phi_{\sigma i}$,

$$\varrho_\sigma(r) = \sum_i n_{\sigma i}\phi^*_{\sigma i}(r)\phi_{\sigma i}(r) \tag{33}$$

and $\varrho(r)$ is given by Eq. (31).

Now, the minimization of the total energy with respect to the spin-orbitals, subject to the orthonormality constraints, leads in this case to one-electron equation of the form

$$\left\{-\frac{1}{2}\nabla^2 - \frac{Z}{r} + \int \frac{\varrho(r_2)}{|r - r_2|} dr_2 + v^\sigma_{xc}[\varrho_\sigma(r)]\right\}\phi_{\sigma i} = \varepsilon_{\sigma i}\phi_{\sigma i} \tag{34}$$

Now one has two sets of one-electron equations (one for spin-up and one for spin-down), which have to be solved self-consistently to determine $\varepsilon_{\sigma i}$ and $\phi_{\sigma i}$. Thus, in the spin-polarized case, the whole calculation is duplicated, but the prediction of many properties is greatly improved.

Since spin-polarization effects are very important for many properties, one could try to obtain such information from the spin-non-polarized results, in order to avoid the duplication of the calculation. Thus, one should try to establish a relation between ϱ_σ, and the total charge density of the spin-non-polarized case. The approximate knowledge of ϱ_σ would lead to an approximate prediction of atomic and molecular properties including the spin-polarization effects.

Now, since the spin-non-polarized description amounts to the assumption that there are the same number of spin-up and spin-down electrons, one could use such electronic configuration as the starting point to reach the actual configuration that distinguishes between the number of electrons of a given spin. This means that one could propose a Taylor expansion of $\varrho_\sigma(r)$ around the charge density of the spin-non-polarized case. Thus we write[1]

$$\begin{aligned}\varrho_\uparrow(r) &= \frac{1}{2}\,\varrho(r)\Big|_{N_\uparrow = N_\downarrow} + \frac{1}{2}(N_\uparrow - N_\downarrow)\,\frac{\partial\varrho(r)}{\partial N}\Big|_{N_\uparrow = N_\downarrow} \\ &\quad + \frac{1}{8}(N_\uparrow - N_\downarrow)^2\,\frac{\partial^2\varrho(r)}{\partial N^2}\Big|_{N_\uparrow = N_\downarrow} + \cdots\end{aligned} \tag{35}$$

and

$$\begin{aligned}\varrho_\downarrow(r) &= \frac{1}{2}\,\varrho(r)\Big|_{N_\uparrow = N_\downarrow} - \frac{1}{2}(N_\uparrow - N_\downarrow)\,\frac{\partial\varrho(r)}{\partial N}\Big|_{N_\uparrow = N_\downarrow} \\ &\quad + \frac{1}{8}(N_\uparrow - N_\downarrow)^2\,\frac{\partial^2\varrho(r)}{\partial N^2}\Big|_{N_\uparrow = N_\downarrow} + \cdots\end{aligned} \tag{36}$$

where $N_\uparrow$ and $N_\downarrow$ represent the number of spin-up and spin-down electrons respectively ($N = N_\uparrow + N_\downarrow$). If one knew the charge density, and the derivatives of the charge density with respect to the number of electrons, one could use Eqs. (35) and (36) to estimate $\varrho_\uparrow$ and $\varrho_\downarrow$. From these one could incorporate spin-polarization effects in the description of

1 See page 97

atomic and molecular properties, without having to perform the spin-polarized calculation of the orbitals, and the orbital energies.

There is however a more interesting interpretation of Eqs. (35) and (36). Note that the first derivative appearing in these equations is the Fukui function of a non-spin-polarized calculation (see Eq. (3)). Thus, subtracting these relations one is lead to

$$f(r) = \frac{\varrho_{\uparrow}(r) - \varrho_{\downarrow}(r)}{N_{\uparrow} - N_{\downarrow}} \tag{37}$$

Since the terms containing the second derivative cancel out, Eq. (37) is correct up to second order. But more important than a high degree of accuracy, is the interpretation of the Fukui function in terms of the spin density. Equation (37) provides a quite attractive route to estimate the Fukui function in the molecular case. Instead of having to solve the Kohn-Sham equations for several occupations of the highest molecular orbital and differentiate numerically the charge density as a function of the number of electrons afterwards, one just has to perform a spin-polarized calculation, from which the evaluation of the spin-density is trivial.

In order to test the approximate validity of Eqs. (15) and (37) we have done spin-polarized calculations on several atoms, using the local spin density exchange-correlation functional of Gunnarsson and Lundqvist[32], to be consistent with the spin non-polarized calculations of $f(r)$ and $\phi_H(r)$ described in the previous section. In Figs. (1)–(4) we have plotted the radial distribution function of $f(r)$, $(\varrho_{\uparrow}(r) - \varrho_{\downarrow}(r))/(N_{\uparrow} - N_{\downarrow})$, and $\phi_H^*(r)\phi_H(r)$ (hoao) versus r for Li, N, F, and Na. One may inmediately notice that both reproduce

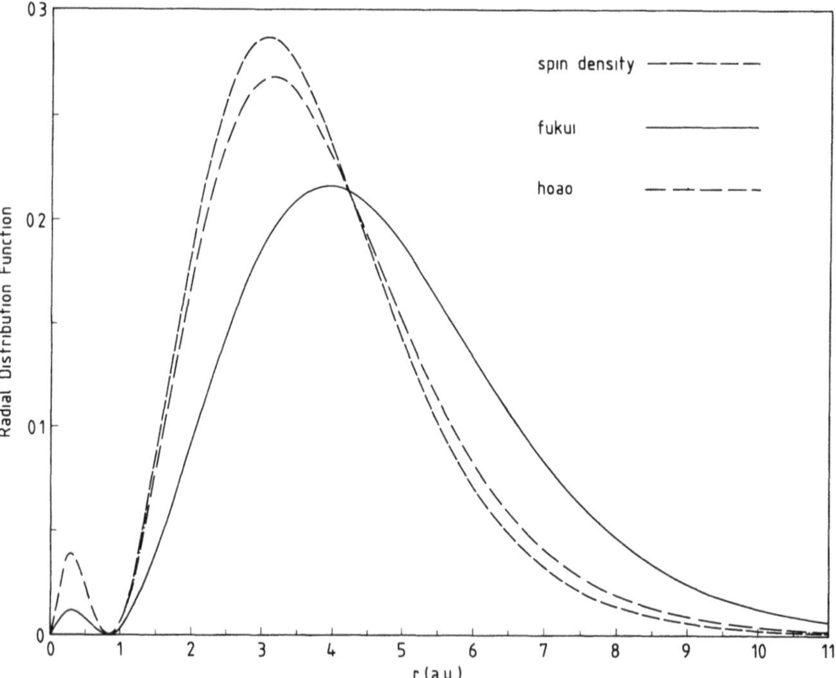

Fig. 1. Li

spin density
fukui
hoao
Radial Distribution Function
0 7
0 6
0 5
0 4
0 3
0 2
0 1
0
0
1
2
3
4
5
6
7
8
9
10
r (a u)

Fig. 2. N

spin density
fukui
hoao
Radial Distribution Function
1 0
0 9
0 8
0 7
0 6
0 5
0 4
0 3
0 2
0 1
0
0
1
2
3
4
5
6
7
8
9
10
r (a u)

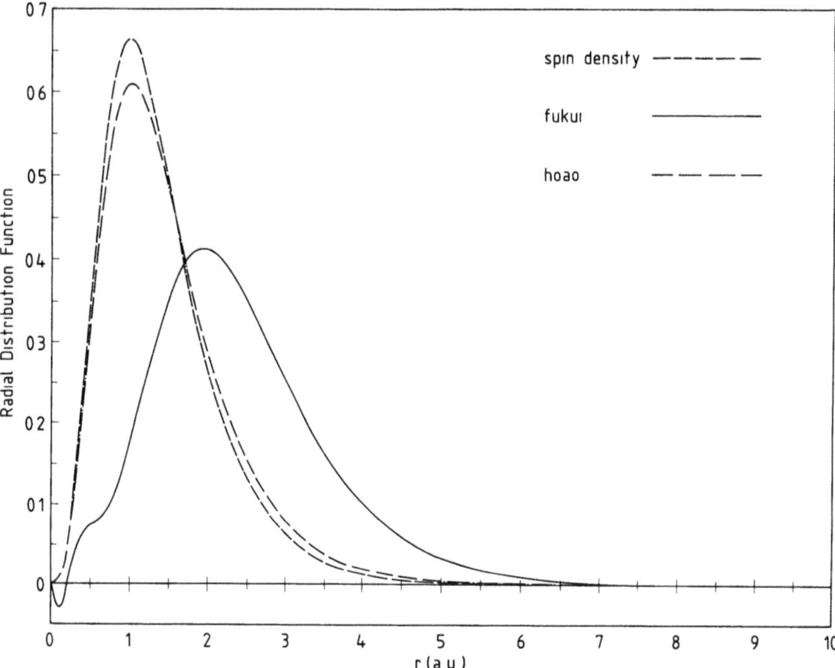

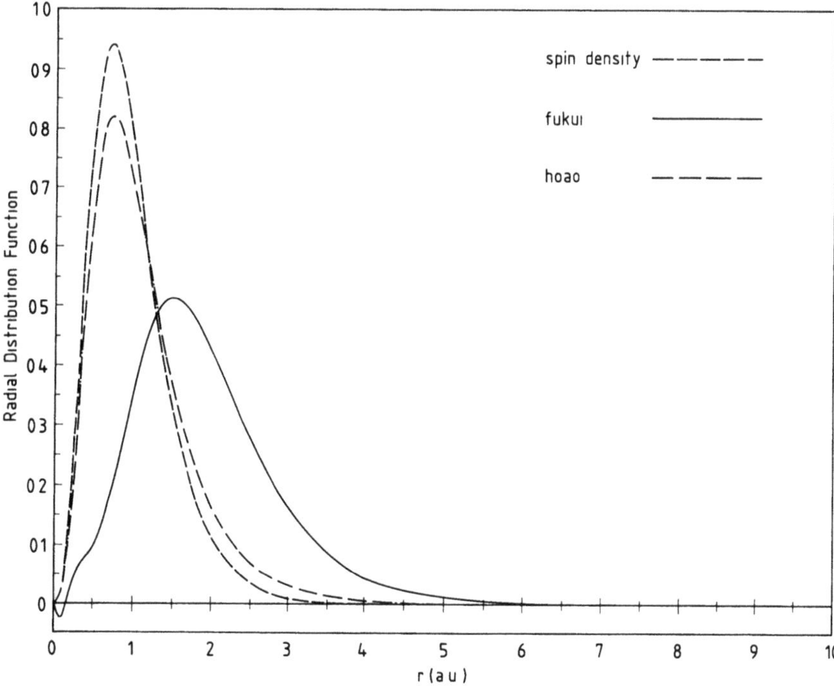

Fig. 3. F

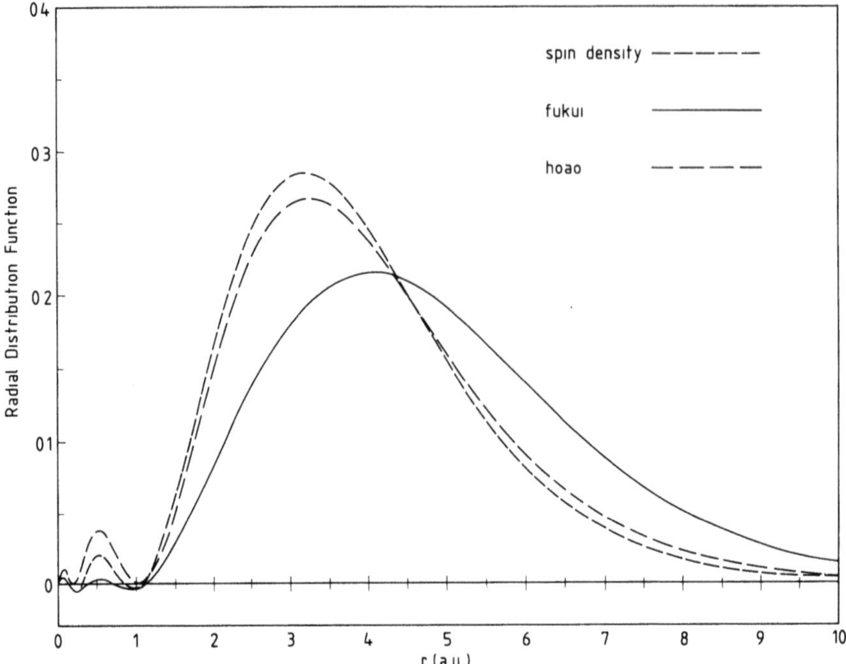

Fig. 4. Na

qualitatively the Fukui function at intermediate and large distances from the nucleus. The structure of $f(r)$ close to the nucleus is not reproduced by either of them, the density of the highest occupied atomic orbital is unable to account for the negative values, and although the spin-density becomes negative at certain points, it does not provide a good description of the innermost region. On the other hand, it may be seen that both provide a reasonable description of $f(r)$ in the outer region, specially for those atoms whose valence orbitals are very diffuse, like Li and Na. Such local behavior has been observed in all open shell atoms that we have studied.

It is interesting to note that the negative values of $f(r)$ lie in the inner regions of the atom, while the large positive values lie in the intermediate and outer regions of the atom.

This observation leads one to conjecture that the local negative values of the Fukui function correspond to non-reactive sites (the atomic core), while local large positive values correspond to reactive sites (the atomic valence region).

Another interesting observation comes from the results described in Figs. 1–4. The spin-density and the highest occupied atomic orbital of open shell atoms, obtained from a spin non-polarized calculation are very similar. Although close to the nucleus they may have significant differences, they become practically identical in the valence region, and therefore it seems that one could include spin-density effects in the outer regions through the knowledge of the density produced by the highest occupied orbital.

It is also interesting to note that if one adds Eqs. (35) and (36), one obtains that

$$[\varrho_\uparrow(r) + \varrho_\downarrow(r)] - \varrho(r)\Big|_{N_\uparrow = N_\downarrow} \cong \frac{1}{4}(N_\uparrow - N_\downarrow)^2 \frac{\partial^2 \varrho(r)}{\partial N^2}\Big|_{N_\uparrow = N_\downarrow} \tag{38}$$

which implies that the difference between the spin-polarized total charge density, and the spin non-polarized total charge density is approximately of second order. An interesting application of Eq. (38) would be to calculate the second derivative of the charge density with respect to the number of electrons from the difference between $\varrho(r)$ spin-polarized and $\varrho(r)$ spin non-polarized. Since the second derivative measures basically the response of the valence orbital to a change in the number of electrons, it may turn out to be an important quantity to describe chemical reactivity.

Now we will consider the same type of expansions made for the charge density, but in relation to the eigenvalues $\varepsilon_{\sigma i}$. Thus, we can write[2]

$$\varepsilon_{\uparrow H} = \varepsilon_H\Big|_{N_\uparrow = N_\downarrow} + \frac{1}{2}(N_\uparrow - N_\downarrow)\frac{\partial \varepsilon_H}{\partial N}\Big|_{N_\uparrow = N_\downarrow} + \frac{1}{8}(N_\uparrow - N_\downarrow)^2\frac{\partial^2 \varepsilon_H}{\partial N^2}\Big|_{N_\uparrow = N_\downarrow} + \cdots \quad (39)$$

and

$$\varepsilon_{\downarrow H} = \varepsilon_H\Big|_{N_\uparrow = N_\downarrow} - \frac{1}{2}(N_\uparrow - N_\downarrow)\frac{\partial \varepsilon_H}{\partial N}\Big|_{N_\uparrow = N_\downarrow} + \frac{1}{8}(N_\uparrow - N_\downarrow)^2\frac{\partial^2 \varepsilon_H}{\partial N^2}\Big|_{N_\uparrow = N_\downarrow} + \cdots \quad (40)$$

According to Eqs. (2) and (12) the first derivative is related to the hardness. Thus if one subtracts these two equations, one obtains

$$\varepsilon_{\uparrow H} - \varepsilon_{\downarrow H} = 2(N_\uparrow - N_\downarrow)\eta \quad (41)$$

where η has been obtained from a non-spin-polarized calculation, and again the terms containing the second derivative cancel out.

As before, Eq. (41) can be used in two directions. That is, the knowledge of the spin-up and spin-down eigenvalues, obtained from a spin-polarized Kohn-Sham calculation, allows one to calculate the hardness. On the other hand, the knowledge of the hardness of an atom allows one to estimate the energy required to flip the spin of the highest occupied atomic orbital if one makes use of the transition state method[30)].

The approximate linearity described by Eq. (41) can be verified in Fig. 5, where we have plotted the difference between the spin-up and spin-down eigenvalues versus the hardness given in Table 3 for halogen atoms, whose value of $N_\uparrow - N_\downarrow = 1$ remains constant. Thus, although the slope is smaller than the value predicted by Eq. (41), the linear relationship is quite reasonably obeyed.

It is interesting to analyze Eq. (41) in relation to the spin flip energy of the elements with the configuration $4s^1 3d^x$. The experimental values are plotted in Fig. 6. It may be seen that they increase almost linearly from K to Cr, and then decrease again almost linearly from Cr to Cu. Now, from the point of view of the Kohn-Sham theory, the spin flip energy can be calculated directly by means of the transition state method[30)]. In order to do that, one would have to perform a self-consistent calculation with half occupancy in the $4s_\uparrow$ and $4s_\downarrow$ orbitals, and the rest of the valence electrons should be accomodated in the $3d_\uparrow$ and $3d_\downarrow$ orbitals according to Hund's rule. Thus, starting from K, one would have $4s_\uparrow^{1/2}\, 4s_\downarrow^{1/2}\, 3d_\uparrow^0\, 3d_\downarrow^0$, and therefore $N_\uparrow - N_\downarrow = 0$ which leads to the correct value of zero for the spin flip energy. From Ca to Cr, one would have $4s_\uparrow^{1/2}\, 4s_\downarrow^{1/2}\, 3d_\uparrow^{N_\uparrow}\, 3d_\downarrow^0$ where $N_\uparrow$ would

2 See footnote 1

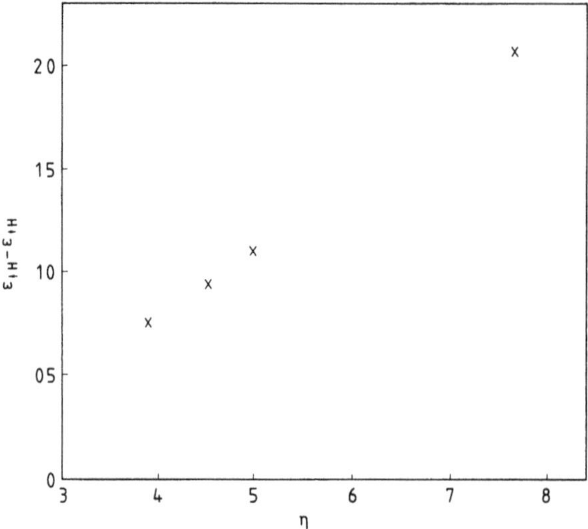

Fig. 5. Difference between the spin-up and spin-down eigenvalues versus hardness for the halogen atoms, $(N_\uparrow - N_\downarrow) = 1$. All quantities in eV

increase from 1 to 5, and therefore $N_\uparrow - N_\downarrow$ would increase linearly from 1 to 5. This analysis shows that if hardness remains approximately constant, then the spin flip energy will increase linearly from K to Cr which agrees with the experimental evidence. After Cr, one begins to fill the $3\,d_\downarrow$ orbital, and therefore $N_\uparrow - N_\downarrow$ will decrease linearly from 5 to zero from Cr to Cu, and for an approximately constant hardness, the spin flip energy would decrease linearly for Cr to Cu, again in agreement with the experimental evidence. This analysis comes in support of Eq. (41) since the variation of the hardness along the transition metal elements varies very slowly[12].

Now, if one adds Eqs. (39) and (40) we find that

$$[\varepsilon_{\uparrow H} + \varepsilon_{\downarrow H}] - 2\,\varepsilon_H\Big|_{N_\uparrow = N_\downarrow} \cong \frac{1}{4}\,(N_\uparrow - N_\downarrow)^2\,\frac{\partial^2 \varepsilon_H}{\partial N^2}\Big|_{N_\uparrow = N_\downarrow} \qquad (42)$$

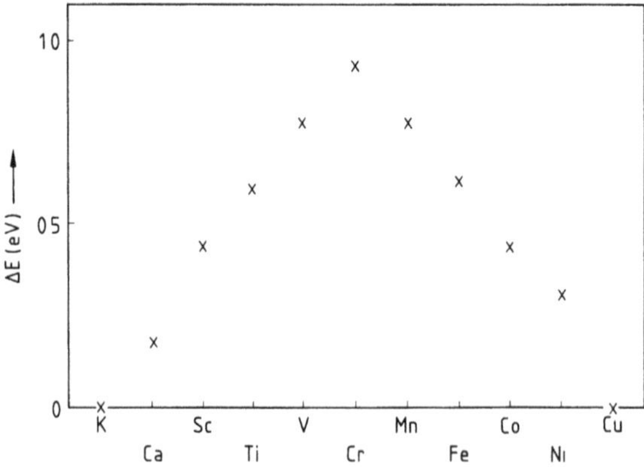

Fig. 6. Experimental spin-flip energy. Values taken from Ref. 44

which indicates that the third derivative of the energy with respect to the number of electrons (see Eq. 12) is approximately given by the difference between the eigenvalues of spin-polarized and a spin-non-polarized calculations. Thus, this derivative, which could be important for estimating the amount of charge transferred in an electronegativity equalization procedure[41], can be estimated in a very simple way in the Kohn-Sham formalism.

Before closing this section, we would like to indicate that the relations derived from the Taylor expansion are also valid for any unrestricted and restricted methods. Therefore, they provide a simple way to evaluate Fukui functions and the hardness of atoms and molecules in the Hartree-Fock theory, or in semiempirical theories.

IV. Chemical Reactivity

We have already mentioned that the definition of the Fukui function has allowed to prove[25] the frontier orbital theory[26] through the use of Eqs. (4) and (5). The results of Parr and Yang indicate that the local values of $f(r)$ should provide information about the reactivity at that site.

In view of relations derived in the previous section we can rewrite Eqs. (4) and (5) in the form

$$dE = \mu dN + \int [\varrho_\uparrow(r) + \varrho_\downarrow(r)] dv(r) dr \tag{43}$$

and

$$d\mu \cong 2\eta dN + \int \left[\frac{\varrho_\uparrow(r) - \varrho_\downarrow(r)}{N_\uparrow - N_\downarrow} \right] dv(r) dr \tag{44}$$

The terms μdN and $2\eta dN$ refer to global changes when there is a net charge transferred in the process of formation of a chemical bond. However the integrands do describe local behaviour. Thus, one can see that the results we have obtained, imply that while energy changes are governed by the total charge density, chemical potential changes are governed by the spin-density. Such an interpretation is quite interesting in that it takes one back to the basic variable of density functional theory, the charge density. Only now it has been refined by breaking it into its spin components. On the other hand, the identification of the Fukui function in terms of the spin density is quite useful to explain some aspects of chemical reactivity.

Let us consider first the formation of a homonuclear diatomic molecule. For this case the global term $2\eta dN$ makes no contribution, thus the initial chemical potential gradient at large distances comes from the spin density. This is in agreement with the facts; for example[42], when two hydrogen atoms are separated by about twice the equilibrium distance, they present a spin polarization, which gives rise to a non-zero spin density. The spin polarization in this case is essential for obtaining the correct free-atom limit. It is important to note that this situation is quite common whenever the molecule dissociates into neutral open shell atoms.

Another important example of spin-density as an indicator of chemical reactivity is provided by the free radicals. These species are known to be highly reactive, which implies the presence of a large initial chemical potential gradient. This may be linked to a large spin density, since the main characteristic of free radicals is the presence of an unpaired electron.

Finally, it is important to mention that Keller and Varea[43] have shown that any source of spin-density plays a very important role in chemisorption and catalysis. This in turn leads again to the identification of high spin-density with reactivity.

Thus, one can see that although the Fukui function and the spin-density may have large differences, from a qualitative point of view they are quite similar in the sense that they provide information of the local reactivity of a species. On the other hand, the close resemblence between the spin-density and the highest occupied orbital implies somehow that the frontier orbital theory could be reinterpreted in terms of spin-density.

Now, in connection with Eq. (41), the proportionality between hardness and the spin flip energy may be used to analyze high and low spin transition metal complexes within the framework of crystal field theory. For the same ligand, soft metals will favour low-spin, while hard metals will favour high-spin.

V. Conclusion

In this chapter we have analyzed some aspects of electronegativity, hardness and the Fukui function from the point of view of the Kohn-Sham formalism. The main results shown are that:

(i) This approach provides a unified and useful procedure to calculate, to inter-relate, and to analyze the electronegativity, the hardness, and the Fukui function of atoms and molecules.

(ii) The Taylor series expansion of the spin-up and spin-down densities around the spin non-polarized charge density leads to a very interesting connection between the Fukui function and spin-density. Although there are large differences between them, especially in the inner region of the atom, qualitatively they are quite similar in the valence region, indicating that either of them may serve as a tool to describe chemical reactivity.

(iii) The postulate of Parr and Yang[25] that states that chemical reactions are preferred from the direction which produces the maximum initial chemical-potential response of a reactant, which implies in turn that local values of $f(r)$ measure reactivity, is strengthened by the interpretation in terms of spin-density. It is known that large spin-density gradients give rise to a high reactivity, for example the free radicals. But large spin-density also implies large chemical potential gradients, Eq. (44).

(iv) The results shown here indicate a close resemblence between spin-density and the highest occupied atomic orbital. However, in order to confirm such behavior, one should study the molecular case and the atomic case in which there is more than one orbital with unpaired electrons, because such systems may show a different behavior.

(v) The Taylor expansion of the spin-up and spin-down Lagrange multipliers around the spin non-polarized eigenvalue leads to a linear relation between the spin-flip energy and the hardness. This relation may prove to be useful for understanding some properties of transition-metal complexes.

Acknowledgment. We wish to thank the Departamento de Cómputo of the Universidad Autónoma Metropolitana-Iztapalapa and Raúl Mejía for their valuable assistance.

Footnote: The Taylor series described by Eqs. (35) and (36) are obtained by neglecting the variation of the charge density of a given spin with respect to the number of electrons of the other spin. The same approximation is used in the case of the eigenvalues to obtain Eqs. (39) and (40).

VI. References

1. Parr, R. G., Donnelly, R. A., Levy, M., Palke, W. E.: J. Chem. Phys. *68,* 3801 (1978)
2. Parr, R. G., Bartolotti, L. J.: J. Phys. Chem. *87,* 2810 (1983)
3. Donnelly, R. A., Parr, R. G.: J. Chem. Phys. *69,* 4431 (1978)
4. Donnelly, R. A.: ibid. *71,* 2874 (1979)
5. Ray, N. K., Samuels, L., Parr, R. G.: ibid. *70,* 3680 (1979)
6. Bartolotti, L. J., Gadre, S. R., Parr, R.: J. Am. Chem. Soc. *102,* 2945 (1980)
7. Sen, K. D., Schmidt, P. C., Weiss, A.: Theor. Chim. Acta *58,* L69 (1980)
8. Sen, K. D., Schmidt, P. C., Weiss, A.: J. Chem. Phys. *75,* 1037 (1981)
9. Gopinathan, M. S., Whitehead, M. A.: Isr. J. Chem. *19,* 209 (1980)
10. Parr, R. G., Pearson, R. G.: J. Am. Chem. Soc. *105,* 7512 (1983)
11. Böhm, M. C., Sen, K. D., Schmidt, P. C.: Chem. Phys. Lett. *78,* 357 (1981)
12. Robles, J., Bartolotti, L. J.: J. Am. Chem. Soc. *106,* 3723 (1984)
13. Gázquez, J. L., Ortíz, E.: J. Chem. Phys. *81,* 2741 (1984)
14. Nalewajski, R. F., Parr, R. G.: ibid. *77,* 399 (1982)
15. Nalewajski, R. F., Capitani, J. F.: ibid. *77,* 2514 (1982)
16. Perdew, J. P., Parr, R. G., Levy, M., Balduz Jr., J. L.: Phys. Rev. Lett. *49,* 1691 (1982)
17. Nalewajski, R. F.: J. Chem. Phys. *78,* 6112 (1983)
18. Politzer, P., Parr, R. G., Murphy, D. R.: ibid. *79,* 3859 (1983)
19. Parr, R. G., Bartolotti, L. J.: J. Am. Chem. Soc. *104,* 3801 (1982)
20. Katriel, J., Parr, R. G., Nyden, M. R.: J. Chem. Phys. *74,* 2397 (1981)
21. Ray, N. K., Parr, R. G.: ibid. *73,* 1334 (1980)
22. Parr, R. G.: Ann. Rev. Phys. Chem. *34,* 631 (1983)
23. Nalewajski, R. F.: J. Am. Chem. Soc. *106,* 944 (1984)
24. Nalewajski, R. F., Koniński, M.: J. Phys. Chem. *88,* 6234 (1984)
25. Parr, R. G., Yang, W.: J. Am. Chem. Soc. *106,* 4049 (1984)
26. Fukui, K.: Theory of Orientation and Stereoselection, Springer-Verlag, West Berlin 1973
27. Pritchard, H. O., Summer, F. H.: Proc. R. Soc. London Ser. *A235,* 136 (1956);
    Iczkowski, R. P., Margrave, J. L.: J. Am. Chem. Soc. *83,* 3547 (1961);
    Klopman, G.: J. Chem. Phys. *43,* 124 (1965)
28. Yang, W., Parr, R. G., Pucci, R.: ibid. *81,* 2862 (1984)
29. Kohn, W., Sham, L. J.: Phys. Rev. *140,* A1133 (1965)
30. Slater, J. C.: Quantum theory of Molecules and Solids, Vol. 4, McGraw-Hill, New York 1974
31. Hohenberg, P., Kohn, W.: Phys. Rev. *136,* B864 (1964)
32. Gunnarsson, O., Lundqvist, B. I.: ibid. *B13,* 4274 (1976)
33. Janak, J. F.: ibid. *B18,* 7165 (1978)
34. Vela, A., Galván, M., Gázquez, J. L.: to be published
35. Galván, M., Vela, A., Gázquez, J. L.: to be published
36. Pariser, R.: J. Chem. Phys. *21,* 568 (1953)
37. Pariser, R., Parr, R. G.: ibid. *21,* 466 (1953)
38. Pople, J. A., Beveridge, D. L.: Approximate Molecular Orbital Theory, McGraw-Hill, New York 1970
39. McGlynn, S. P., Vanquickenborne, L. G., Kinoshita, M., Carroll, D. G.: Introduction to Applied Quantum Chemistry, Holt, Rinehart and Winston Inc., New York 1972
40. Keller, J.: Density functional theory, Lecture Notes in Physics, Vol. 187, 1 (1983)
41. Sanderson, R. T.: Science *121,* 207 (1955)
42. Gaŕritz, A., Gázquez, J. L., Castro, M., Keller, J.: Int. J. Quantum Chem. *15,* 731 (1979)
43. Keller, J., Varea-Alvarez, C.: ibid. *12,* Suppl. 1, 165 (1977)
44. Harris, J., Jones, R. O.: J. Chem. Phys. *70,* 830 (1979)

Electronegativity of Atoms and Molecular Fragments

K. D. Sen[1], M. C. Böhm[2] and P. C. Schmidt[2]

1 School of Chemistry, University of Hyderabad, Hyderabad – 500134, India
2 Institut für Physikalische Chemie, Physikalische Chemie III, Technische Hochschule Darmstadt, 6100 Darmstadt, F.R.G.

Theoretical calculations of electronegativity χ for atoms and molecular fragments carried out by the authors using the Slater Transition State and the Transition Operator concept have been reviewed. The difference between such calculations which yield the estimates of χ as per the Mulliken definition and the actual density functional theoretical definition is highlighted by performing several calculations of energy as a function of occupation numbers in atoms within the framework of an approximate local density approach. The theoretically derived χ values have been compared with other empirical/semiempirical estimates and have been found to rationalize the latter in general. Reliable predictions on the charge distribution and related properties in molecules can be made using these theoretical χ values. Limitations of the calculations on molecular orbital electronegativity are discussed.

A. Introduction 100

B. Method of Calculations 102
B.I. Key Parameters 102
B.II. Atoms 102
B.III. Molecules 103

C. Results and Discussion 104
C.I. Atoms 104
C.II. Molecules 112

D. Future Scope of Work 121

E. References 122

Structure and Bonding 66

A. Introduction

The concept of electronegativity, χ, first introduced by Pauling[1] is based on chemical intuition. Its definition in terms of the "power" of an atom in a molecule to attract electrons towards itself has led to several scales of electronegativity which depend on a given choice of measuring the "power". Mulliken[2] defined χ as an arithmetic mean of ionization potential, I, and the electron affinity, A, thereby making spectroscopic and/or orbital estimates of χ feasible. Such a definition regards χ as a well defined one-electron property of an atom in a given state. A review of the work done along these definitions is given by Mullay[3]. The spectroscopic approach based on the atomic line spectral data on several atoms in the ground as well as excited states to define χ in a given valence state has been particularly reviewed by Bergmann and Hinze[4]. For an N electron system with total energy, E, and one-electron density, $\varrho(r)$, the definition proposed by Pritchard and Sumner[5a] as well as by Iczkowski and Margrave[5b], though lacking in a sound theoretical basis as originally proposed, can now be understood in the light of the density functional theory[6,7], namely dE/dN as given in Eq. (1) below. The latter result worked out by Parr and coworkers[8] has opened up exciting possibilities of calculating χ for N electron systems (atoms, molecules, clusters) Given the exact energy density functional $E[\varrho]$, the *foremost task* is that of testing this approach in its ability to rationalize the previously obtained empirical/semiempirical results on χ. In absence of knowledge about the exact density functional $E[\varrho]$, only the available approximations to it have to be used! Bartolotti et al.[9] have presented a review of similar studies based on the Kohn-Sham density functional. Other simple density functionals used in relation to χ have been reviewed by Alonso et al.[10].

The Mulliken electronegativity χ can be calculated in a straightforward manner within the conventional Hartree-Fock (HF) Self-Consistent Field (SCF) procedure[11], via the ΔSCF method in which two separate calculations of E defining the process $M^+ \rightarrow M^-$ with $\Delta N = 2$ are carried out and ΔE is calculated in terms of the difference in the E values[12]. As compared to the HF-SCF theoretical definitions of the occupation number n_i and the one-electron eigenvalue ε_i corresponding to the i-th orbital, the local density approximation (LDA) to the density functional ascribes an entirely different significance to these quantities[7]. In the latter, n_i is allowed to change continuously from 0 to 1 and ε_i represents the negative derivative of E with respect to the occupation number n_i. This new interpretation of ε_i leads to the concept of the Slater transition state (TS)[13] which due to its computational simplicity provides a reliable alternative to the ΔSCF method. With reference to the calculations of χ based on the TS approach, χ^{TS}, within a given LDA, a single SCF calculation of ε_i, and not E, has to be carried out on a transition state (see below). In addition the E value is estimated directly in terms of the ε_i's thus avoiding the numerical errors involved in taking the difference between two large quantities as $E_2 - E_1$ in the ΔSCF method. However, it is also possible to implement the ΔSCF method with a given LDA to calculate χ as χ_0^{LDA}. Such values of χ_0^{LDA} are more accurate than χ^{TS} as the former include complete electronic relaxation effects. Since the numerical errors of ΔSCF calculations are not severe for atomic systems, in the outer valence region the two approaches of calculating χ^{TS} and χ_0^{LDA}, give results of comparable accuracy. It is important to remember in this context that these approaches at best provide estimates of χ which correspond to the finite difference approximation (see Eq. (3) below) and do not actually represent χ within the density functional theory (Eq. (1) below). In addition, the level of accuracy of the calculated value of χ also depends on the details of the LDA actually employed in the calculations. Variants of LDA exist in the form of the Slater X_α[13], Kohn-Sham[6b], Gunnarson-Lundquist[14], Vosko et al.[15] etc. exchange-correlation potential, respectively. The accuracy of the calculations can be further improved by incorporating corrections due to spin-polarization, self-interaction and relativistic effects, respectively.

For molecules the computational advantages of the TS approach can be realized within the transition operator (TO) approach[16] developed within the framework of the LCAO-MO approximation. This model is particularly well suited for carrying out the calculations of χ, χ^{MO}, on molecular fragments specifically defined with respect to a given chemical bond in the parent molecule. In this type of calculations the chemical bond Y $\cdots$ Z is studied in terms of the two molecular fragments Y and Z, respectively, frozen in the geometry as in the parent molecule and considering a suitable electronic state of the fragment closely resembling the valence state. A comparison of the χ^{MO} corresponding to the two fragments can be made and the nature of the associated chemical bond can be discussed. The accuracy of calculations depends on the details of the LCAO-MO model used. The above-mentioned possibilities of carrying out quantum chemical calculations of χ do not at all aim at providing yet another set of scales of χ. Instead, they simply represent the stepwise modifications leading to a consistent quantum mechanical approach that would enable chemists to calculate χ for an N electron system given in a well defined geometry and electronic state.

Similar to the concept of electronegativity, chemists would also prefer to have a quantitative theory of yet another very useful principle, i.e. that of hard and soft acids and bases first introduced by Pearson[17]. Several empirical scales of hardness and softness are also available[18]. Recently Parr and Pearson[19] have given a theoretical justification to this principle and have quantified the same in terms of the absolute hardness parameter, η (see Eq. (2) below). The general discussion on the quantum chemical calculations of χ also holds good in the context of η. In this chapter we shall present a review of the calculations carried out by the present authors for a large number of atoms[20–24] and molecular fragments[22–27] with special reference to the *basic task* set forth by the work of Parr et al.[8] as already mentioned above. We note here that on one hand a set of atomic χ values can be used in conjunction with the principle of equalization of electronegativity (PEE)[28, 29] to obtain molecular group electronegativities. On the other hand the direct method of the TO approach also provides results for the same quantity. The extent to which it is possible to rationalize these two sets of results is directly related to the question of how far one is able to develop a transparent theory. This is another aspect of the *basic task* which will be considered at the end of this review. The following discussion will be carried out in the context of the model-dependent nature of these calculations, e.g. the validity of the one-electron approximation as well as effects of covalent and Coulombic interactions on the charge separation in real systems.

In Sect. B, we have presented the definition of the key parameters and briefly illustrated the methods of calculation adopted in our studies[20–27] on χ and η. In Sect. C, we have compiled these results and discussed their relevance as purely quantum chemical parameters in predicting experimental properties such as electron affinity, atomic charges, nuclear quadrupole coupling constants etc. In Sect. D the scope of future work along these lines is given.

B. Method of Calculations

B.I. Key Parameters

The density functional theoretical definition of electronegativity χ and absolute hardness parameter η is given by

$$\chi = -(dE/dN)_v \tag{1}$$

and

$$\eta = 0.5\,(d^2E/dN^2)_v \tag{2}$$

respectively, where v defines the external potential. The definitions given above are actually employed within a certain LDA by performing several calculations of E as a function of N. An approximation to Eq. (1) and (2) can be given in terms of the finite difference form according to

$$\chi \approx 0.5\,(I + A) \tag{3}$$

and

$$\eta \approx 0.5\,(I - A) \tag{4}$$

respectively. Such approximate definitions have been popularly employed within the transition state/operator methods which will be discussed in Parts II and III in the following.

B.II. Atoms

Within the nonrelativistic spin-polarized X_α-theory[13)] the one-electron equations are given by

$$[f_1 + V_c(r) + V_i^{ex}(r)]u_i(r) = \varepsilon_i u_i(r) \tag{5}$$

where $f_1 = -\nabla^2 - 2\,Z/r$, the Coulomb potential $V_c(r) = \sum_i n_i \int u_i^*(r_2)/|\vec{r} - \vec{r}_2|d\tau_2$ and the exchange potential

$$V_i^{ex}(r) = -\alpha\left(\frac{3}{4}\pi\right)^{1/3} \varrho^{1/3}(r) \tag{6}$$

respectively. As mentioned in the Introduction the one-electron eigenvalues ε_i of Eq. (5) have the following physical meaning

$$\frac{\partial E}{\partial n_i} = \varepsilon_i \tag{7}$$

Equation (7) forms the basis of the concept of the Slater transition state. If the initial and final electronic states, respectively, are denoted by a set of occupation numbers $(\ldots, n_i, n_j, \ldots)$ and $(\ldots, n_i', n_j', \ldots)$ in the one-electron orbitals, i, j, ..., the corresponding TS is defined by the set of occupation numbers $[0.5\,(n_i + n_i'), 0.5\,(n_j + n_j'), \ldots]$. The difference in energy, ΔE, corresponding to the electronic transition can be calculated as

$$\Delta E \approx \sum_i (n_i - n_i')\, \varepsilon_i^{tr} \tag{8}$$

where the ε_i^{tr} refer to the one-electron eigenvalues obtained via a single SCF calculation carried out on the TS. When a unique orbital is involved in the electron detachment (ionization) as well as the attachment (affinity) process the TS configuration corresponding to the calculation of χ corresponds to precisely that of the ground state atom. Li $(1s^2 2s^1)$, B $(1s^2 2s^2 2p^1)$, ... form the examples of this category of atoms wherein the 2s and 2p orbitals uniquely define the ionization and affinity processes. When different one-electronic states define the active orbitals for these two processes, for example in Be $(1s^2 2s^2)$ and a majority of atoms corresponding to the d- and f-block elements in the Periodic Table, the TS configuration must be suitably defined. In the calculations of ΔE according to Eq. (8) two eigenvalues from a single SCF calculation have to be employed in such cases. Successful applications of Eq. (8) in predicting ΔE for a variety of electronic transitions in atoms have been reported by Sen et al.[30]. Recently, a rigorous theoretical basis for the concept of TS has been discussed[31]. The χ values obtained from TS calculations at best provide approximations to the density functional estimate in Eq. (1) and give values for the Mulliken definition as given in Eq. (3). The same is true of the hardness parameter η. Slightly more accurate values of similar nature can be obtained via ΔSCF calculations with $\Delta N = 2$ within the same LDA. The direct use of Eq. (1) and (2) for χ and η, respectively, by using an accurate functional representation of E would in principle lead to the best theoretical estimates within the same LDA. Böhm and Schmidt[24] have recently carried out similar calculations of χ^{LDA} and η^{LDA} by expressing E as a polynomial function of two independent parameters defining hybridization and charge transfer. Such a set of atomic χ values then have been combined with the PEE to obtain estimates of molecular group electronegativities. By virtue of being the most consistent theoretical calculation within LDA we shall compare the results of these calculations with those based on the TS approach in Sect. C.

B.III. Molecules

Analogous to the TS a transition operator (TO) can be defined [16] within the HF LCAO-MO theory. The quantities of interest namely, I and A corresponding to the active orbital can be directly calculated in terms of the one-electron eigenvalues of a suitably defined Fock operator. Thus, a molecular fragment with doublet spin multiplicity in the ground state having a surplus of one α spin electron would be defined by the transition Fock operator

$$F(1)_{i\alpha} = F(1)_\alpha - 0.5\, \langle i^t(2) \| i^t(2) \rangle \tag{9}$$

for the ionization process. The corresponding operator for the electron attachment process would be described as

$$F(1)_{i\beta} = F(1)_\beta + 0.5 \langle i^t(2) \| i^t(2) \rangle \quad (10)$$

The one-electron eigenvalues ε_i in

$$F(1)_{i\alpha}|i^t(1)\rangle = \varepsilon_i^t|i^t(1)\rangle \quad (11)$$

and

$$F(1)_{i\beta}|i^t(1)\rangle = \varepsilon_i^t|i^t(1)\rangle \quad (12)$$

directly provide the values of I and A, respectively. The TO approach defined in Eq. (9)–(12) allows for electronic relaxations accompanying ionization and electron attachment. A detailed analysis of relaxation energies within the TO formalism has been discussed in the literature[32]. The TO calculations of χ^{MO} have been performed for a large number of molecular fragments using specifically developed CNDO and INDO-MO Hamiltonians[33]. The latter employ dressed two-electron integrals to reproduce experimental ionization potentials in the outer valence region. For molecular calculations, in this manner a consistency in the treatment of the two-electron integrals has been incorporated. A general limitation of the TO approach ought to be kept in mind. The validity of the one-particle approximation is a basic assumption in this theory. While in several organic fragments the condition of a single active orbital for χ is fulfilled, in transition metal fragments it is usually not the case. In the former situation the calculations of χ and η can be correlated with the concept of frontier orbitals. This topic has been treated by Gázquez et al.[34]. The breakdown of the single orbital picture and its repurcussions on the prediction of experimental values of nuclear quadrupole coupling constants in special cases will be taken up in Sect. C.

Lastly, we refer to the interesting possibility of estimating the molecular group electronegativities and hardness using a set of empirical or semiempirical atomic χ and η values. This topic is the subject matter of the chapter by Mortier[35] as well as Bergmann and Hinze[4]. A critical comparison of such calculations along with those based on the TO approach with the LDA calculations of Böhm and Schmidt[24] will be made in the next section.

C. Results and Discussion

C.I. Atoms

Bartolotti et al.[36] have reported the first calculations of χ for the 54 atoms corresponding to the main group elements of the Periodic Table using the TS approach within the X_α theory. Sen et al.[20] reported similar calculations of χ for the rare earth atoms with special reference to the attachment of an electron in the 4f shell using the relativistic X_α theory[37]. In Table 1 the results of χ for rare earth atoms are reproduced. The numerical results follow the well-known trend of near constancy of χ within this series. In order to extract the values of A indirectly from χ, separate TS calculations of I corresponding to the process of electron detachment (as defined for the calculations of χ) were also carried out. Such values of I along with the estimates of A have been included in Table 1. The various TS defining the calculations of χ and I have also been shown. The estimates of A in Table 1 indicate that mononegative ions of rare earth atoms would be unstable in the gas phase. The calculations of A reported by other authors[38)] refer to the electron attachment process in the 5d orbital and the stability of the mononegative ions predicted for some of the rare earth atoms refer to entirely different electronic configurations than considered here. The TS calculations of χ (denoted as χ^{TS}) for 36 atoms in the main group of the Periodic Table have been reported by Schmidt and Böhm[22)] using the relativistic LDA due to Vosko et al.[15)]. These results are presented in Table 2, column 3. Separate calculations of I and A were also carried out[22)] (Table 1). The calculated values of A were found to be in excellent agreement with the experimental estimates due to Hotop and Lineberger[39)]. The scaled values of χ (i.e. χ^{scale}) according to

$$\chi^{scale} = 0.19\,(I + A - 1) \tag{13}$$

were also summarized in order to make comparisons with the spectrum of classical scales of χ. Such a comparison shows that χ^{TS} values are in general good agreement with the previous empirical data on atomic χ values in those cases where the valence shell configuration is given by s^1, s^2 or s^2p^5. Deviations are found in the cases of configuration s^2p^1 – s^2p^4. This observation emphasizes the need for carrying out accurate calculations of χ on the valence states. We shall once again refer to the χ^{TS} values given in Table 2 for the analysis of the nuclear quadrupole coupling constant data for diatomic molecules. Böhm and Schmidt[24)] have recently reported more accurate calculations of χ^{LDA} for a large number of atoms in the ground and valence states by directly using Eq. (1) within the LDA due to Vosko et al.[15)]. These authors have expressed χ as a function of the partial

Table 1. Values of the first ionization potential, electronegativity, and electron affinity in eV for the rare earth atoms. The two j values of 5/2 and 7/2 are denoted by – and + as the orbital subscripts, respectively. For the 5d orbital 5d- denotes $5d_{j=3/2}$. The atoms Yb and Lu show deviations due to the participation of the 5d orbital instead of the 4f[20)]

Atom	Ionization potential: I		Electronegativity: χ^{TS}		Electron affinity: A
	Transition state	Transition energy	Transition state	Transition energy	
Ce	$4f_-^{2.5}6s^1$	8.73	$4f_-^{3}\ 6s^1$	1.80	−5.13
Pr	$4f_-^{3.5}6s^1$	8.36	$4f_-^{4}\ 6s^1$	1.84	−4.86
Nd	$4f_-^{4.5}6s^1$	8.10	$4f_-^{5}\ 6s^1$	1.90	−4.30
Pm	$4f_-^{5.5}6s^1$	7.89	$4f_-^{6}\ 6s^1$	1.93	−4.03
Sm	$4f_+^{0.5}6s^1$	8.24	$4f_+^{1}\ 6s^1$	1.96	−4.32
Eu	$4f_+^{1.5}6s^1$	8.17	$4f_+^{2}\ 6s^1$	1.95	−4.27
Gd	$4f_+^{2.5}6s^1$	8.16	$4f_+^{3}\ 6s^1$	2.02	−4.12
Tb	$4f_+^{3.5}6s^1$	8.15	$4f_+^{4}\ 6s^1$	2.00	−4.15
Dy	$4f_+^{4.5}6s^1$	8.27	$4f_+^{5}\ 6s^1$	2.01	−4.25
Ho	$4f_+^{5.5}6s^1$	8.39	$4f_+^{6}\ 6s^1$	2.03	−4.33
Er	$4f_+^{6.5}6s^1$	8.49	$4f_+^{7}\ 6s^1$	2.04	−4.41
Tm	$4f_+^{7.5}6s^1$	8.72	$4f_+^{8}\ 6s^1$	2.06	−4.60
Yb	$6s^{1.5}$	6.01	$5d_-^{0.5}6s^{1.5}$	1.85	−2.31
Lu	$5d_-^{1.5}6s^1$	2.45	$5d_-^{2}\ 6s^1$	3.87	+5.29

Table 2. The calculated values of the electronegativity χ (in eV) and the absolute hardness parameter η (in εV) for atoms using the TS method via Eq. (3) and (4) and the LDA via Eq. (1) and (2). Ground state (gs) and valence state (vs) are defined by the electron configuraton[22, 24]

Atom	gs	χ^{TS}	χ^{LDA}	vs	χ^{LDA}	Atom	gs	η^{TS}	η^{LDA}	vs	η^{LDA}
Li	s	3.14	2.93	sp	2.31	Li	s	2.52	1.72	sp	1.68
Be	s	4.52	6.26	sp	4.60	Be	s	4.53	2,78	sp	2.74
B	p	4.60	4.41	sp^2	6.37	B	p	3.88	3.80	sp^2	3.86
C	p	6.72	5.67	sp^3	8.03	C	p	4.95	4.83	sp^3	4.86
N	p	7.58	7.70	sp^3	10.51	N	p	7.34	5.75	sp^3	5.73
O	p	7.93	9.57	sp^3	13.25	O	p	5.83	6.62	sp^3	6.68
F	p	11.02	11.47	sp^3	16.16	F	p	6.92	7.73	sp^3	7.75
Na	s	3.06	2.90	sp	2.25	Na	s	2.44	1.52	sp	1.49
Mg	s	3.86	5.89	sp	4.21	Mg	s	3.86	2.26	sp	2.18
Al	p	3.14	3.68	sp^2	5.35	Al	p	2.82	2.82	sp^2	2.90
Si	p	4.92	4.80	sp^3	6.53	Si	p	3.33	3.44	sp^3	3.50
P	p	5.76	6.09	sp^3	8.25	P	p	4.80	4.01	sp^3	4.08
S	p	6.44	7.45	sp^3	10.08	S	p	4.03	4.49	sp^3	4.57
Cl	p	8.48	8.79	sp^3	11.92	Cl	p	4.64	5.32	sp^3	5.36
K	s	2.60	2.98	sp	2.48	K	s	2.03	1.70	sp	1.57
Ca	s	3.11	2.98	sp	3.46	Ca	s	3.08	1.93	sp	1.87
Ga	p	3.30	3.44	sp^2	6.00	Ga	p	2.77	2.95	sp^2	2.99
Ge	p	4.68	4.83	sp^3	6.83	Ge	p	3.23	3.44	sp^3	3.51
As	p	5.48	5.94	sp^3	8.29	As	p	4.35	3.86	sp^3	3.93
Se	p	6.15	7.10	sp^3	9.10	Se	p	3.93	4.24	sp^3	4.31
Br	p	7.69	8.15	sp^3	11.30	Br	p	4.18	4.86	sp^3	4.90
Rb	s	2.53	2.41	sp	1.93	Rb	s	1.94	1.33	sp	1.29
Sr	s	2.96	4.37	sp	3.28	Sr	s	2.89	1.81	sp	1.81
In	p	3.21	3.63	sp^2	5.63	In	p	2.69	2.66	sp^2	2.77
Sn	p	4.30	4.98	sp^3	6.83	Sn	p	2.89	3.07	sp^3	3.39
Sb	p	5.15	5.58	sp^3	7.54	Sb	p	3.66	3.49	sp^3	3.55
Te	p	5.78	6.61	sp^3	8.83	Te	p	3.74	3.78	sp^3	3.84
I	p	6.77	7.18	sp^3	10.06	I	p	3.61	4.19	sp^3	4.24
Cs	s	2.36	2.76	sp	2.33	Cs	s	1.80	1.28	sp	1.24
Ba	s	2.74	3.94	sp	3.01	Ba	s	2.62	1.81	sp	1.81
Tl	p	3.37	4.02	sp^2	6.14	Tl	p	2.93	2.95	sp^2	3.04
Pb	p	3.95	4.86	sp^3	6.73	Pb	p	3.48	3.28	sp^3	3.34
Bi	p	4.30	5.79	sp^3	7.90	Bi	p	4.30	3.55	sp^3	3.62
Po	p	5.10	6.72	sp^3	9.08	Po	p	3.61	3.78	sp^3	3.85
At	p	5.85	7.59	sp^3	10.26	At	p	3.33	4.23	sp^3	4.27

atomic charge, q, and the hybridization parameter, r, in the valence shell configuration $s^r p^{1-r}$ according to the relationship

$$\chi(q, r) = A_0 + B_0 r + (A_1 + B_1 r)q + (A_2 + B_2 r)q^2 + \cdots \tag{14}$$

In Table 3 we have listed the set of fitting parameters (A_i, B_i) in Eq. (14). The theoretical values χ (denoted by χ^{LDA}) corresponding to the ground and valence states under χ_{gs}^{LDA} and χ_{vs}^{LDA} have been listed in Table 2 in column 4 and 6, respectively. A comparison of the values in the valence states with the experimental estimates in the cases of $s^2p^1 - s^2p^4$ corrects for the anomaly reported in the discussion on the estimates of electronegativity based on the TS approach as given in Table 2, column 3. The ground

Table 3. Coefficients for the dependence of the electronegativity on the s character r_a of the chemical bond and the partial charge q_a of the atom. The electronegativity is given by $\chi_{a,r}(r_a, q_a) = A_{0,a} + B_{0,a}r_a + (A_{1,a} + B_{1,a}r_a)q_a + (A_{2,a} + B_{2,a}r_a)q_a^2$. The data are given for all main group elements besides the rare gases and hydrogen. For hydrogen one gets for pure s electronic states $\chi_a = 6.55 + 10.93\ q_a$ [24)]

	A_0	B_0	A_1	B_1	A_2	B_2
Li	1.68	1.25	3.27	0.16	0.68	1.65
Be	2.94	3.32	5.31	0.32	1.60	0.76
B	4.41	5.56	7.60	0.15	1.51	0.72
C	5.97	8.22	9.65	0.29	1.76	0.04
N	7.70	11.24	11.49	−0.12	2.11	0.65
O	9.57	14.72	13.24	0.47	2.59	−0.91
F	11.47	18.74	15.45	0.17	2.56	1.32
Na	1.60	1.30	2.85	0.18	0.11	2.85
Mg	2.53	3.36	4.19	0.33	1.00	0.46
Al	3.68	5.10	5.64	0.46	0.04	0.67
Si	4.80	6.91	6.88	0.49	0.77	0.09
P	6.09	8.65	8.02	0.50	1.14	−0.21
S	7.45	10.52	8.98	0.61	1.85	−0.95
Cl	8.79	12.51	10.63	0.31	1.74	1.72
K	1.47	1.01	2.86	0.54	0.46	2.97
Ca	2.29	2.33	3.62	0.23	0.50	0.42
Ga	3.84	6.48	5.89	0.58	−0.05	0.93
Ge	4.83	7.99	6.87	0.56	0.65	0.13
As	5.94	9.40	7.71	0.56	1.11	−0.22
Se	7.10	10.89	8.47	0.56	1.63	−0.81
Br	8.15	12.60	9.71	0.35	1.65	−1.84
Rb	1.44	0.97	2.49	0.17	−0.02	2.11
Sr	2.18	2.19	3.60	0.02	−0.04	0.80
In	3.63	6.01	5.32	0.63	−0.34	1.10
Sn	4.48	7.38	6.14	0.55	0.45	0.10
Sb	5.58	7.83	6.98	0.49	0.79	−0.38
Te	6.61	8.87	7.55	0.49	1.30	−0.73
I	7.18	11.51	8.37	0.39	1.51	1.84
Cs	1.39	0.87	2.41	0.14	−0.09	1.86
Ba	2.08	1.86	3.39	−0.06	−0.01	0.75
Tl	4.02	6.36	5.89	0.54	−0.17	0.94
Pb	4.86	7.46	6.55	0.53	0.46	0.08
Bi	5.79	8.43	7.10	0.50	0.87	−0.19
Po	6.72	9.45	7.55	0.54	1.40	−0.80
At	7.59	10.66	8.46	0.33	1.50	−1.66

state values of χ^{TS} are found to be smaller than the χ^{LDA} values. The χ_{vs}^{LDA} values are found to be significantly different from the χ_{gs}^{LDA} values. This can be understood in terms of the changes brought about due to hybridization in the magnitude of I and A, respectively, which are of similar order and contribute additively to χ. We note here that the similar variations in hybridization is not expected to change the magnitude of the absolute hardness parameter η as the individual effects on I and A would tend to cancel out in the approximate definition of η given by Eq. (4). In Table 4 the χ_{vs}^{LDA} results have been compared with those of Bergmann and Hinze[4)] for a few atomic species. As mentioned in the Introduction the latter set of values are based on an entirely different, spectroscopic

Table 4. Comparison of valence state electronegativities (in eV) calculated by two different methods[24)]

Atom	Valence state		Bergmann Hinze[a]	Böhm Schmidt[b]
H	s		7.18	6.55
Be	$(sp)^2$		4.79	4.60
	$(sp)^1\pi^1$	sp	4.47	4.60
		π	3.23	2.94
B	$(sp^2)^3$		6.31	6.26
	$(sp)^1\pi^2$	sp	7.08	6.69
		π	5.13	4.41
C	$(sp^3)^4$		8.06	8.02
	$(sp)^2\pi^2$	sp	10.75	10.08
		π	6.43	5.97
Si	$(sp^3)^4$		7.26	6.53
	$(sp)^2\pi^2$	sp	9.03	8.26
		π	5.65	4.80

[a] Bergmann and Hinze, Ref. 4; [b] Böhm and Schmidt, Ref. 24

Table 5. The calculated electronegativity χ^{TS} in eV, for the ground and excited state electronic configurations of Li, Be and B. The SCF criterion on the eigenvalue has been set to 10^{-5} Ryd. Ground-state values are marked with $^+$

Atom	Configuration	Transition state	χ^{TS}
Li	2s	$2s^{0\,5,0\,5}$	2.58^+
	3s	$3s^{0\,5,0\,5}$	1.00
	4s	$4s^{0\,5,0\,5}$	0.60
	2p	2p	1.80
	2p	$2p^{0\,5,0\,5}$	1.30
	3p	3p	0.99
	3p	$3p^{0\,5,0\,5}$	0.74
	4p	4p	0.60
	4p	$4p^{0\,5,0\,5}$	0.45
	3d	3d	0.75
	4d	4d	0.53
Be	2s	$2s^{0\,5}2p^{0\,5}$	3.48^+
	2s2p	2s2p	3.25
	2s3p	2s3p	1.25
	2s4p	2s4p	0.71
	2s3d	2s3d	0.71
	2s4d	2s4d	0.55
B	2p	2p	4.08^+
	2p	$2p^{0\,5,0\,5}$	3.40
	3p	3p	1.34
	3p	$3p^{0\,5,0\,5}$	0.99
	4p	4p	0.74
	4p	$4p^{0\,5,0\,5}$	0.56
	3s	3s	1.98
	4s	4s	0.95

approach. The degree of quantitative agreement found in the two sets of electronegativity results in Table 4 is gratifying. Sen[21] has further carried out the calculations of χ^{TS} using the spin-unrestricted X_α-theory for several atomic Rydberg states. These results are given in Table 5 for Li, Be, and B, respectively. A significantly large decrease in the magnitude of electronegativity is found in going from the ground state to the first excited state which can be explained on the basis of the changes in the s or p character of the active orbitals. The changes in characteristic orbital properties are not very significant within the various Rydberg states of the same atom as a result of which the χ^{TS} values are also not changing drastically within the set of excited states. A general conclusion on the basis of the above calculations is that the χ corresponding to the ground state is numerically found to be the maximum. A rigorous theoretical proof of this result would be a useful analysis in the density functional theory. The valence states must be considered separately due to their virtual nature. With a proper hybridization it is possible to obtain χ values in either direction of the ground state value. We conclude the discussion on the atomic calculations by summarizing in Fig. 1 how these values compare with the other experimental estimates of χ. In Fig. 1, the spectrum of empirical estimates of χ have been displayed for atoms in a given periode. The estimates of χ^{TS} due to Schmidt and Böhm[22] are given with the numerical scattering resulting from the various other scales. In Fig. 2 we have shown the values of electronegativity (a) for the ground state using the TS approach, (b) for the valence state using the LDA along with Eq. (1), and (c) the empirical estimates as given in Fig. 1 due to Allred and Rochow[3]. The atoms are arranged along a given group in the Periodic Table. The correct ordering of the elements in IIIA group vis-a-vis those in the IIA group as well as the ordering of elements within group IIIA itself is indicative of the importance of valence states in defining χ for atoms. A similar conclusion is obtained while locating the position of the oxygen atom in relation to those of fluorine and chlorine which is correctly obtained via the valence state calcula-

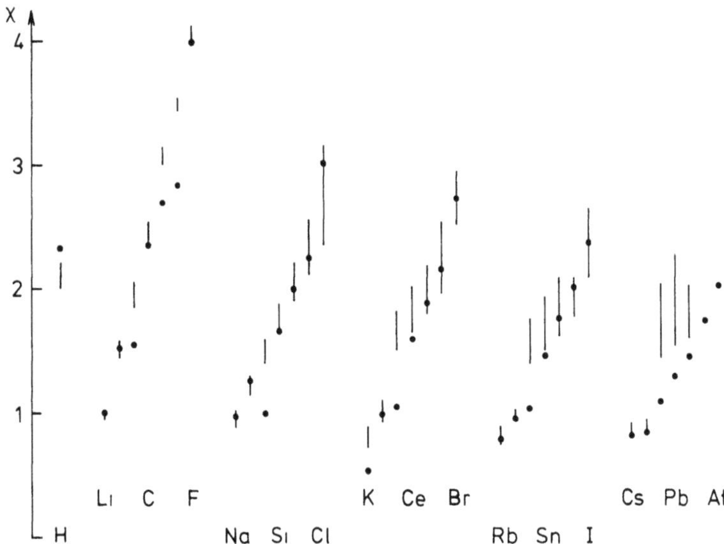

Fig. 1. The electronegativity values for atoms as given by the various scales. Ground state values due to Schmidt and Böhm[22] are shown as *circles*

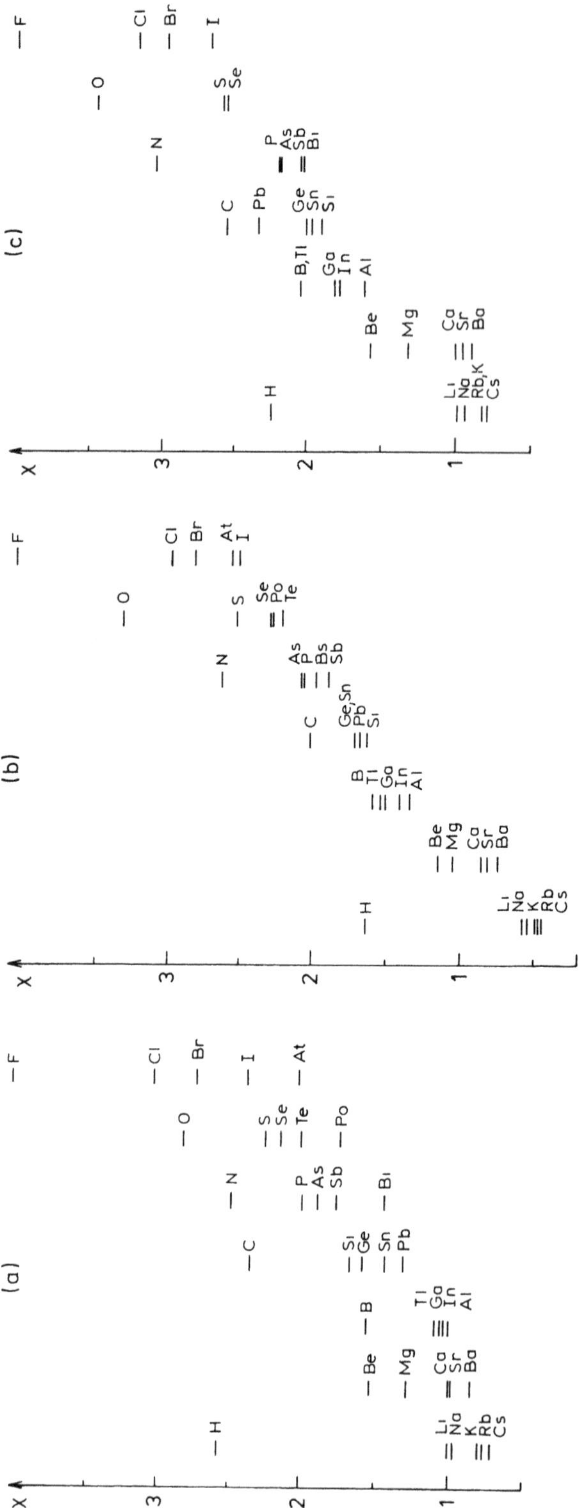

Fig. 2a–c. The electronegativity values for atoms in the ground state as calculated via the transition state approach **(a)**, in the valence state as calculated via the direct use of Eq. (1) **(b)**, and the experimental estimates due to Allred and Rochow as in Fig. 1 **(c)**

tion of χ. The calculations of the absolute hardness parameter, η, which forms a companion parameter to χ within the density functional theory, can be performed analogously. As in the case of χ, the TS calculations of η must be regarded as only approximations to those results which are based on the direct use of Eq. (2). Nevertheless, its finite difference counterpart according to Eq. (4) forms a useful guideline in following certain qualitative trends in η. Any erroneous conclusion based on such an analysis must, however, be reanalyzed in terms of Eq. (2). Sen et al.[23] have calculated the values of η based on the TS approach for several atoms in the ground state. These values are reproduced in Table 2, column 9. A comparison with the experimental trends suggests that the calculated values are in general good agreement with experiments except in the cases of the elements of the V group of the Periodic Table. The more accurate calculations of Böhm and Schmidt[24] based on the direct application of Eq. (2) reproduce the expected trends within a given row of the Periodic Table. These η values are collected in column 10 of Table 2. In column 11 of the same table the results on several valence states are also given. As discussed earlier the changes in the values of η due to hybridization are much less pronounced than in χ. Due to the dominant values of I the qualitative trends in both χ and η are found to be similar with respect to the changes in hybridization. Next we consider the correlation of χ and η with another important structural property of atoms namely the static dipole polarizability, α_d. Assuming that a larger polarizability can be associated with softer atoms one would expect that η values correlate well with the inverse of polarizability, $(\alpha_d)^{-1}$, where α_d represents the spherically averaged static polarizability. As shown in Fig. 3, for a large number of atoms, an excellent linear correlation has been found between η and an independent calculation of α_d^{-1} based on the

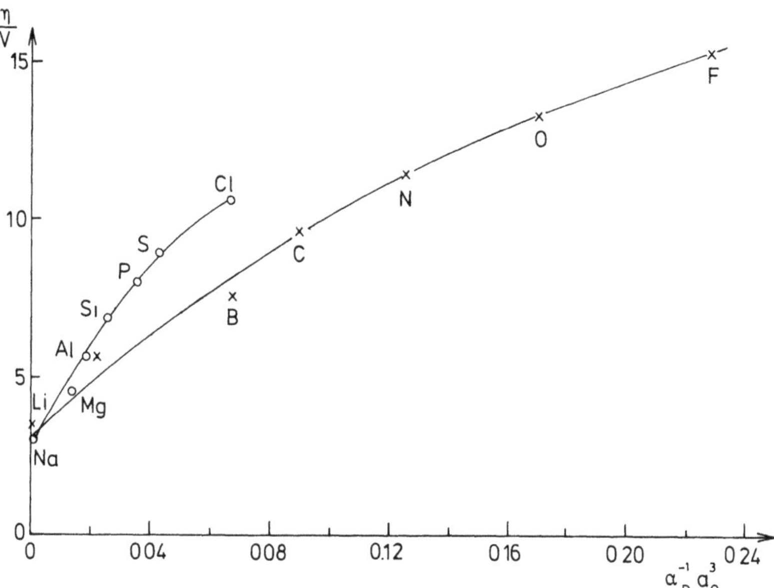

Fig. 3. The variation of the absolute hardness parameter and the averaged dipole polarizability tor atoms. All values have been calculated using the same LDA. The values of η refer to the definition given in Eq. (2) in the text and not the approximate relationship given in Eq. (4)

same LDA[40]. In this connection it must be mentioned here that an orbitalwise resolution of α_d for the ground and valence states of a very large number of atoms[40] leads to the important conclusion that the dependance of $(\alpha_d)^{-1}$ on hybridization is qualitatively similar to the one found on χ and η, respectively. The significance of the recent work of Shanker and Parr[41], wherein a satisfactory account of structure-stability properties for a large number of binary compounds has been obtained on the basis of χ and η values, should be viewed in the light of above correlation.

C.II. Molecules

The molecular group electronegativity, usually derived by means of atomic spectral properties has been reported previously by Jaffe and Hinze[42], Huheey[43], and more recently by Bergmann and Hinze[4], respectively. We shall denote these estimates of electronegativity by χ_{gp}. Calculations using purely theoretical atomic electronegativity, χ^{LDA}, have also been carried out[24] along with the PEE and various estimates of group electronegativity, χ_{gp}^{LDA}, have been made. On the other hand, in the spirit of the TS calculations of χ for atoms, its related LCAO counterpart namely the TO approach has been applied to the molecular fragments as defined in Sect. II and the molecular orbital electronegativity values, χ^{MO}, have been calculated[22, 24–27] within the LCAO-MO theory. In this section we shall present a review of χ^{MO} calculations and then compare its performance vis-a-vis the other χ_{gp} values.

Böhm et al.[25] carried out the calculations of χ^{MO} for a number of organic fragments containing Be, B, C, N, and O as the central atom, respectively. A modified CNDO-MO Hamiltonian[33] with standard molecular geometry was used throughout these calculations. A significantly useful feature of such χ^{MO} calculations lies in the possibility of carrying out a quantitative analysis of the variation of electronegativity with the percentage atomic character, X%, in the active molecular orbital wave function. Thus valuable chemical insight in terms of electron delocalization effects on χ^{MO} is gained from such a theoretical approach. In Table 6, we have listed the results of χ^{MO} calculations along with the relevant values of X% in the active MO corresponding to the ionization and affinity processes, respectively. The separate values of the calculated I and A are also given. It is found that X% can be reduced up to 40%. Therefore one can expect that the predictions based on the atomic electronegativity corresponding to ground states are unreliable for the molecules. From Table 6 it is clear that a strong localization (large X%) leads to an enhancement in χ^{MO} in molecules with central atoms from the right side of the Periodic Table. The compensating effect of the electron delocalization in the active MO vis-a-vis the change in the central atom in the molecular fragments can be discussed by considering the example of CCl_3/BCl_2 and CBr_3/BBr_2, respectively. A stronger localization of the active MO in the B fragments proportionately restricts the decrease in χ^{MO} as would have been normally expected in going from C to B in the Periodic Table. The results obtained in Table 6 suggest that the useful approach of atoms in molecule can be formulated via the TO calculations of electronegativity.

Schmidt and Böhm[22] have carried out TO calculations of χ^{MO} for the molecular fragments G in molecules of the general type $R_1R_2R_3CCl$ (G: $R_1R_2R_3C$) and $R_1R_2C = CR_3Cl$ (G: $R_1R_2C = CR_3$), respectively. A modified INDO MO Hamiltonian[33] and experimental geometry of GCl molecules in the gas phase have been used throughout the

Table 6. Calculated TO ionization potentials (I), electron affinities (A), electronegativities according to Mulliken (χ^{MO}) and Pauling (χ^{P}) for oxygen, nitrogen, carbon, boron and beryllium fragments according to a semiempirical CNDO Hamiltonian. The X contribution to the corresponding group orbital is indicated (X = Be, B, C, N, O). The values of I, A, and χ^{MO} are given in eV[25]

Molecular fragment	I^{TO} (eV)	% X in I-TO	A^{TO} (eV)	% X in A-TO	χ^{MO}	χ^{P}
BeH	7.42	94.9	0.14	96.0	3.78	1.06
$BeCH_3$	7.74	86.4	1.17	88.2	4.46	1.29
BeF	12.02	98.9	5.12	99.3	8.56	2.67
BeCl	8.31	96.8	1.45	97.7	4.88	1.19
BeBr	7.56	98.0	9.75	98.5	4.16	1.19
BH_2	9.49	92.6	0.85	95.5	5.17	1.53
$B(CH_3)_2$	8.95	76.4	2.08	77.5	5.52	1.65
BF_2	12.48	87.2	4.12	88.0	8.30	2.58
BCl_2	9.83	76.0	2.58	80.2	6.21	1.88
BBr_2	8.33	67.1	1.64	72.7	4.99	1.47
CH_3	10.40	95.6	0.39	97.0	5.40	1.61
CH_2F	11.27	87.1	1.63	86.9	6.45	1.95
CHF_2	12.23	76.9	2.77	80.1	7.50	2.30
CF_3	13.97	72.7	4.58	72.3	9.28	2.91
CH_2Cl	9.95	78.4	1.10	88.5	5.53	1.64
$CHCl_2$	9.75	70.5	1.91	67.2	5.83	1.74
CCl_3	9.74	66.3	2.56	70.4	6.15	1.86
CH_2Br	9.14	64.8	9.74	83.8	4.94	1.44
CHB_2	8.44	58.9	1.36	50.7	4.90	1.43
CBr_3	8.19	54.2	1.86	40.2	5.03	1.48
NH_2	11.81	90.4	0.58	90.1	6.20	1.86
NF_2	16.28	43.2	6.28	65.9	11.28	3.56
NCl_2	11.61	46.8	3.09	60.8	7.35	2.25
NBr_2	10.36	30.5	2.86	47.4	6.61	2.00
OH	16.30	82.0	3.29	78.9	9.80	3.07
OF	18.91	42.5	8.18	39.3	13.55	4.32
OCl	16.01	56.7	5.75	54.9	10.88	3.43
OBr	14.90	58.8	5.16	53.2	10.03	3.14

calculations. The results are reproduced in Table 7. Before discussing the correlation of χ^{MO} values with the ^{35}Cl nuclear quadrupole coupling constant (column 6, Table 7), it is necessary to discuss the limitations of the one-electron approximation, inherent in the present calculations of χ^{MO}. Reliable predictions of bond polarity can only be made through χ^{MO} if the chemical bonding is satisfactorily represented by a unique active orbital from each fragment. In order to illustrate this point we present an analysis of the active MO in the CH_3 and CH_2Cl fragment, respectively. The GCl sigma bond formed between the CH_3 unit and Cl is well represented by a single active orbital. A majority of fragments in Table 7 fall under this category. In Table 8 we have shown the details of the active orbitals in terms of the C character in the latter two fragments (i.e. CH_3 and CH_2Cl). Whereas the analysis for CH_3 corroborates this uniqueness (i.e. the validity of

Table 7. Ionization energies I, electron affinities A, and electronegativities χ^{MO} for some fragments G derived by means of the transition operator (TO) method. The nuclear quadrupole coupling constants $C = e^2qQ/h$ for molecules $G^{35}Cl$ in the gas phase are given too. The values of I, A, and χ^{MO} are given in eV[22)]

No.	Fragment G	I	A	χ^{MO}	$C(G^{35}Cl)$ MHz
1	$HC{\equiv}C{-}CH_2$	11.64	1.40	6.52	−75.80
2	CH_3	12.25	0.38	6.32	−74.74
3	$C_6H_5{-}CH_2$	10.33	2.28	6.31	
4	CH_3CH_2	11.17	−0.07	5.55	−70.28
5	$(CH_3)_3C$	9.94	−0.73	4.61	−67.60
6	CHF_2	12.28	1.09	6.69	
7	$CHCl_2$	13.39	−0.12	6.64	−77.90
8	$CHFCl$	12.84	0.16	6.50	
9	CH_2Cl	12.81	0.01	6.41	−78.40
10	trans-$HClC{=}CH$	12.28	0.78	6.53	−78.70
11	$H_2C{=}CCl$	12.13	0.81	6.47	−78.70
12	$F_2C{=}CF$	12.34	0.10	6.22	−73.70
13	$H_2C{=}CH$	11.89	0.53	6.21	−68.89
14	$HC{\equiv}C{-}CH{=}CH$	11.07	1.32	6.19	−69.80
15	cis-$HClC{=}CH$	11.76	0.50	6.13	−72.30
16	$H_2C{=}CH{-}CCl{=}CH$	10.73	0.94	5.83	−68.10
17	$H_2C{=}CH{-}C{=}CH_2$	10.53	1.08	5.80	−67.60
18	$H_2C{=}CCH_3$	10.58	0.74	5.66	
19	$C{\equiv}N$	14.13	4.26	9.20	−83.27

Table 8. Localization properties (LCAO amplitudes) and orbital energies ε_i (in eV) from one-electron wave functions in the CH_3 and CH_2Cl fragments (=G) that show large C lone-pair admixtures. The two moieties are associated to GCl systems summarized in Table 7. The hypothetical GCl σ-bond lies in the direction of the z axis; thus only the 2s and $2p_z$ coefficients are given. The first fragment is well represented by a single active orbital while this approximation breaks down in the Cl derivative. The ground state is a doublet with a surplus of one α-spin electron[22)]

Fragment	α or β spin	MO no. (i)	ε_i (eV)	$C_{i,2s}$ $C_{i,2p_z}$
CH_3	α	4	−12.25	0.310 0.917
	β	4	−0.38	0.344 0.919
CH_2Cl	α	5	−12.81	0.268 0.660
	α	7	−10.55	0.219 0.580
	β	7	−0.01	0.380 0.850

the orbital picture), the situation is quite different for CH_2Cl. In the latter, the two occupied α-spin orbitals have considerable admixtures from the C lone-pair in the donor level but the acceptor level in the β-spin subspace has a strongly localized C character. The concept of molecular orbital electronegativity is not expected to apply under such situations. In Table 7, the fragments CH_2Cl, $CHCl_2$, and C≡N fall under this category. We now discuss the correlation of the high quality data on the ^{35}Cl quadrupole coupling constant $C(^{35}Cl)$ in gas phase with the calculated χ^{MO} values. In Fig. 4, we have plotted $C(^{35}Cl)$ vs. χ^{MO} values. Strong linear correlations are obtained in the form of two topologically similar sets of fragments defined by sp^2 and sp^3 hybridizations, respectively. Such a linear correlation further highlights the predominance of bond polarity in determining the variation in C values. Note also the sensitivity of χ^{MO} values to the geometry of the molecular fragment, for example, in the cases of cis and trans fragments of HClC = CClH, respectively. Similarly finer aspects like differences in $C(^{35}Cl)$ values corresponding to Cl(α) and Cl(β) positions in $H_2C = CH - CCl_\alpha = CHCl_\beta$ can also be rationalized in terms of χ^{MO} values. The predominance of bond polarity controlling the coupling constant is also exemplified by the linear correlation found between $C(^{35}Cl)/C_0(^{35}Cl)$ vs. $\chi_A^{TS} - \chi_{Cl}^{TS}$ values in a series of molecules of the type AB (B = Cl) as shown in

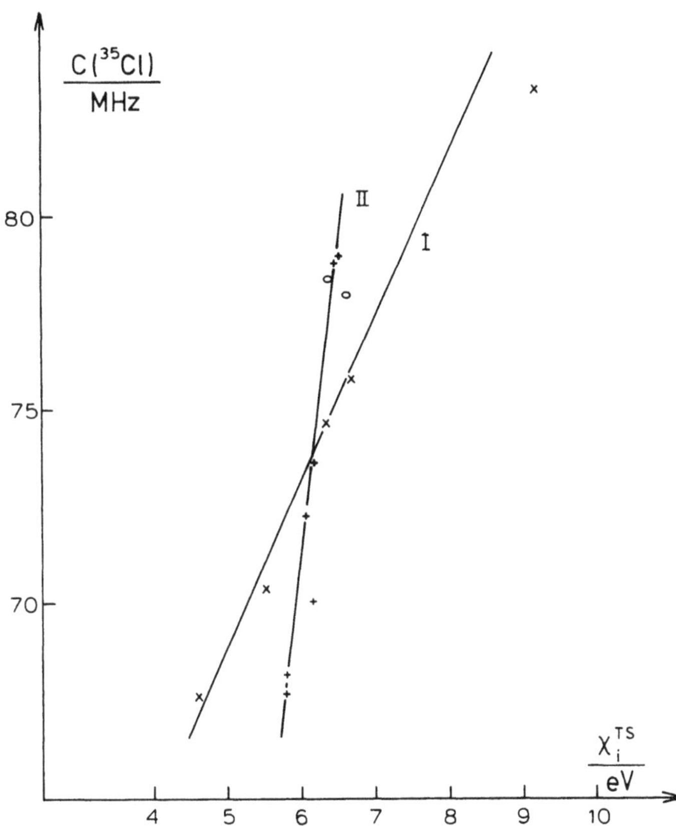

Fig. 4. A correlation of the molecular orbital electronegativity with the gas phase data on the ^{35}Cl quadrupole coupling constant. The linear correlations I and II respectively refer to the molecular fragments involving sp^2 and sp^3 hybridized carbons

Fig. 5. The departure from linearity found in the cases of the fragments of $CHCl_2$, CH_2Cl and C≡N can be traced to the breakdown of the one-electron approximation. In Table 9, χ^{MO} values of a series of some new fragments have been calculated with the purpose of further checking upon the sensitivity of χ^{MO} on modulations of the molecular environment around a C atom. These fragments have been generated by removing the H atom from a CHR group coupled to another organic fragment via a C=C unit in the parent

Table 9. Ionization energies I, electron affinities A and electronegativities χ^{MO} for some organic fragments G derived by means of the transition operator (TO) method[22)]

Fragment G	I	A	χ^{MO}
HC=O	11.20	0.63	5.92
FC=O	11.29	2.65	7.78
HC=S	10.19	1.28	5.74
$HC{=}CHCH_3$	11.22	0.51	5.87
$HC{=}C(CH_3)_2$	11.07	0.45	5.78
$HC{=}CCH_3Cl$	10.58	0.74	5.66
$HC{=}CHSCH_3$	11.14	0.73	5.94
$HC{=}C{=}CH_2$	11.02	1.95	6.49
HC=C=O	11.54	0.35	5.95
HC=CH–CH=CHCl	10.73	0.94	5.84
$HC{=}CH{-}CCl{=}CH_2$	10.53	1.08	5.81

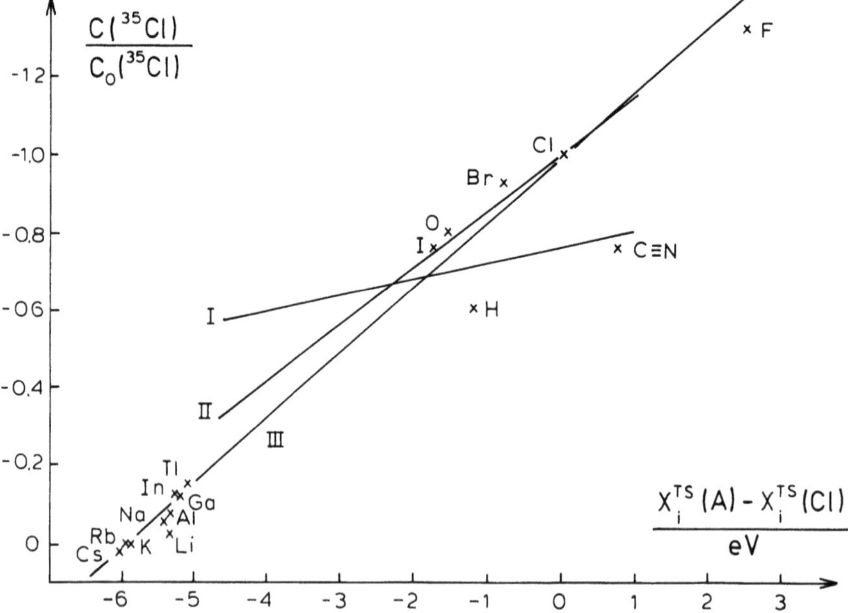

Fig. 5. A correlation of the difference in the atomic group electronegativity values representing a measure of bond polarity in several diatomic molecules with the ^{35}Cl quadrupole coupling data. *Curves I and II* of Fig. 4 have also been shown

molecule. A clear variation in χ^{MO} values in Table 9 suggests that such data would be quite useful in analysing the experimental data on deuterium quadrupole coupling constants when they become available.

The theoretical values can also be used to estimate the partial ionic character[44] (PIC) of the associated chemical bond between the two units. The strong dependence of χ^{MO} values on the electronic state and molecular geometry makes it possible to obtain reliable estimates of ionic vs. covalent character through such calculations of PIC. Using the relationship

$$PIC = 16\,[\chi^{P}_{Li} - \chi^{P}_{fg}] + 3.5\,[\chi^{P}_{Li} - \chi^{P}_{fg}] \tag{15}$$

with $\chi^{P} = 0.336\,(\chi^{Mulliken} - 0.6157)$, Sen et al.[27] have analyzed the nature of C–Li bonds in a series of simple lithiated hydrocarbons. In Table 10 the results of these calculations are reproduced. Due to near degeneracies of different electronic states in the corre-

Table 10. The calculated ionization potentials (I), electron affinities (A), electronegativities χ^{MO} for the molecular fragments derived from lithiated methanes, fluoromethanes, cyclopropane and acetylene (all values in eV). The molecular point groups of the parent compounds (S_p) as well as the point group of the fragment (S_{fg}) are also shown[27]

Molecules	S_p	Fragment	S_{fg}	I	A	χ^{MO}
Li		Li		5.34	0.84	3.09
CH_3Li	C_{3v}	CH_3	C_{3v}	10.40	0.39	5.40
		CH_3	C_{2v}	15.85	6.19	11.02
CH_3Li (planar)	C_{2v}	CH_2Li	C_{2v}	9.31	2.83	6.07
		$CHH'Li$	C_s	9.94	2.82	6.38
CH_2Li_2, tetrahedral, singlet	C_{2v}	CH_2Li	C_s	9.20	3.59	6.40
		$CHLi_2$	C_s	8.43	2.77	5.60
CH_2Li_2, tetrahedral, triplet	C_{2v}	CH_2Li	C_s	8.67	2.81	5.74
		$CHLi_2$	C_s	6.93	2.61	4.77
CH_2Li_2, *cis*-planar singlet	C_{2v}	CH_2Li	C_s	10.37	3.14	6.76
		$CHLi_2$	C_s	7.86	2.95	5.41
CH_2Li_2, *trans*-planar singlet	D_{2h}	CH_2Li	C_{2v}	8.22	1.80	5.01
		$CHLi_2$	C_{2v}	10.21	4.70	7.46
$CHLi_3$	C_{2v}	$CHLi_2$	C_{2v}	10.23	4.75	7.49
		$CHLi'$	C_s	7.48	2.60	5.04
		CLi_3	C_{2v}	7.56	3.53	5.55
		$CHLi_2$	C_s	7.70	2.04	4.87
$CHLi_3$	C_{3v}	CLi_3	C_{3v}	7.44	2.82	5.13
CLi_4	T_d	CLi_3	C_{3v}	7.15	3.04	5.10
	D_{4h}	CLi_3	C_{2v}	7.41	3.44	5.43
CF_2Li_2, tetrahedral	C_{2v}	$CFLi_2$	C_s	6.49	1.37	3.93
		$CFLi$	C_s	7.91	3.15	5.53
CF_2Li_2, *cis*-planar	C_{2v}	CF_2Li	C_s	13.66	7.31	10.49
1,1 dilithiocyclopropane, tetrahedral	C_{2v}	C_3H_4Li	C_s	7.66	1.13	4.40
1,1 dilithiocyclopropane planar	C_{2v}	C_3H_4Li	C_s	10.05	1.50	5.78
HCCLi	$C_{\infty v}$	HCC	$C_{\infty v}$	14.56	3.49	9.03
H_3CCCLi propinyl-lithium	C_{3v}	H_3CCC	C_{3v}	11.94	4.24	8.09

sponding units (i.e. high spin vs. low spin configurations) we have performed various calculations of χ^{MO} in the related model systems. Completely optimized geometry based on sophisticated ab initio LCAO-MO calculations[45] have been used to define the fragments throughout the calculations using an INDO MO formalism[33]. Once again χ^{MO} values have been found that depend sensitively upon the spin multiplicity and molecular geometry. For example χ^{MO} for the methyl fragment changes from 5.4 to 11.02 eV on the Mulliken scale as its symmetry is changed from C_{3V} to C_{2V} point group. A larger electronegativity in the case of cis CH_2Li (planar) over the trans isomer is obtained as a consequence of stronger hybridization effect in the active MO in the cis isomer. This rehybridization is prevented in the trans case. Within a given hydrocarbon successive substitution by Li gradually decreases the χ^{MO} value of the fragment. A comparable magnitude of χ^{TS} for Li and χ^{MO} of the CH_3 fragment in C_{3V} geometry suggests a predominance of covalent character in the C–Li bond (16% ionic character according to Eq. (14)), a conclusion also arrived earlier by Schleyer and coworkers[45]. A comparison of the χ^{MO} values corresponding to several other fragments in Table 10 with χ^{TS} of Li indicates that a range of variation from predominantly covalent to ionic bonding is to be expected in the nature of the C–Li bond depending upon the extent of lithium substitution in the fragment.

Böhm et al. have reported[26] the calculations of χ^{MO} for molecular fragments containing 3d-metal atoms at the central position. The experimental geometry of the fragment as it exists in the parent molecule has been considered and a modified INDO MO model has been employed in the calculations of χ^{MO}. These results are presented in Table 11 (column 7) along with the estimates of the I and A values (column 3 and 5). In each case the analysis of the active MO corresponding to the electron detachment and attachment process has been also given (column 4 and 6). In the last column a set of scaled values

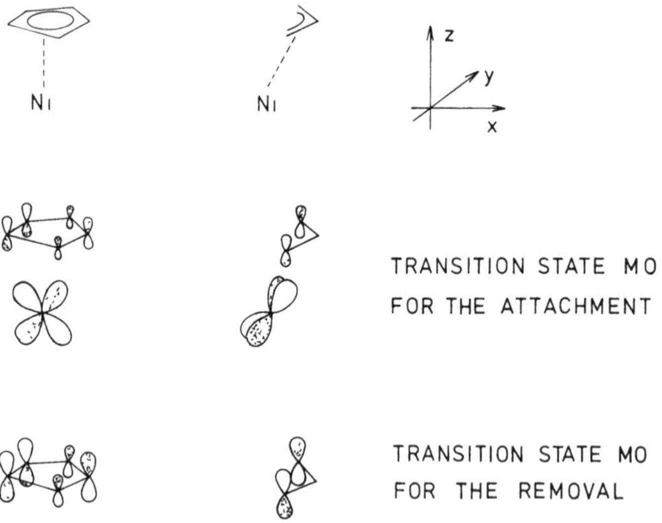

Fig. 6. The representation of the different active molecular orbitals for the process of electron removal and attachment in the examples of the half-sandwich molecular fragments of $Ni(C_3H_3)$ and $Ni(C_5H_5)$, respectively, with the choice of a coordinate-axes system

Table 11. Electronegativity χ^{MO} calculated using the UHF INDO TO method for a series of free radicals containing the 3d transition metal atoms. All values are in eV[26)]

Compound	Multiplicity	I_1 (eV)	MO-type of the i'th cationic transition MO	A_1(eV)	MO-type of the i'th anionic transition MO	χ^{MO}	χ^P
VO_4	quartet	14.47	O 2pπ ($1t_1$)	4.85	O 2p ($1t_1$)	9.66	3.02
CrO_4	triplet	12.75	O 2pπ ($1t_1$)	5.03	O 2p ($1t_1$)	8.89	2.76
MnO_4	doublet	12.22	O 2pπ ($1t_1$)	5.22	O 2p ($1t_1$)	8.72	2.71
$CuCl_4$	doublet	10.18	Cl 3pπ	5.98	Cl 3pπ	8.08	2.49
NiCl	doublet	8.29	Cl 3pσ, Ni 3dσ 93%	−1.19	Ni 3dσ, Cl 3pσ 17%	3.55	0.98
NiBr	doublet	7.59	Br 4pπ 99%	−3.22	Ni 3dσ 97%	2.19	0.52
$Ni(C_5H_5)$	doublet	7.79	Cp(π), Ni $3d_{xz}/3d_{yz}$ 5%	−0.72	Ni $3d_{yz}$, Cp(π) 83%	3.54	0.98
$Ni(C_3H_5)$	doublet	7.04	A(π)*, Ni $3d_{yz}$ 2%	−0.87	Ni $3d_{yz}$. A(π)* 71%	3.09	0.82
Ni(NO)	doublet	6.52	Ni 3dσ, NO 5σ 92%	−2.41	NO(2π)*, Ni 3dπ 4%	2.06	0.48
$Co(CO)_4$	doublet	7.92	Co $1e(3d_{xz}/3d_{yz})$	0.93	Co $3d_{z^2}$, CO(π)*	4.43	1.27
$Fe(C_5H_5)$	doublet	7.87	Cp(π), Fe $3d_{xz}/3d_{yz}$	−0.16	Cp(π)*	3.86	1.08
$Fe(CO)_3$	triplet	8.66	Fe $1e(3d_{x^2-y^2}, 3d_{xy})$	0.14	Fe $2e(3d_{xz}, 3d_{yz})$	4.40	1.26
$Mn(CO)_3$	doublet	8.04	Mn $2e(3d_{xz}, 3d_{yz})$	1.39	Mn $2e(3d_{xz}, 3d_{yz})$	4.72	1.37
$Mn(CO)_5$	doublet	7.67	Mn $3d_{z^2}$ 57%	1.42	Mn $3d_{z^2}$ 18%	4.55	1.31
$Cr(CO)_3$	singlet	8.81	Cr $1a_1(3d_{z^2})$	1.19	Cr $2e(3d_{xz}/3d_{yz})$	5.00	1.46
$Cr(C_6H_6)$	singlet	6.13	Cr $3d_{z^2}$	0.29	Bz(π)*, Cr $2e(3d_{x^2-y^2}/3d_{xy})$	3.21	0.87

according to Eq. (13) are given in order to facilitate comparisons with other empirical scales. As obtained in columns 4 and 6 the majority of fragments are characterized by the nonconservation of the active MO in the processes leading ionization and electron attachment. As a consequence, pronounced effects of reorganization of electronic charge distribution are to be expected. A schematic display of the donor and acceptor (active) orbitals in the cases of the half sandwich fragments $Ni(C_3H_3)$ and $Ni(C_5H_5)$, respectively, is given in Fig. 6 along with the choice of the coordinate axes system. For the cyclopentadienyl system, 4.8% Ni 3d character is found for the donor function while the acceptor function is localized to the extent of 83.5% at the transition metal center. For the allyl fragment the corresponding values are 2.3% and 71.1%, respectively. The largest values for χ^{MO} are found in the set VO_4, CrO_4, MnO_4 and $CuCl_4$. These values are in the same range as found for the electron deficient fragments in Table 6. This is as per the expectations that the singlet ground state ions of the type VO_4^{2-} are generally found for such polyatomic species. The calculated χ^{MO} for the five fragments containing Ni are much smaller. Among the two halides, NiBr corresponds to a lower value indicating a relatively larger donor capability than NiCl. As a result of the electron-rich centre, e.g. Ni in comparison to the main group elements, e.g. Be, the variation in χ^{MO} values for NiCl to NiBr is much more pronounced than in the corresponding BeCl to BeBr fragments. The low values of χ^{MO} for the half-sandwich Ni fragments signify enhanced donor properties as is exemplified by the formation of multiple decker sandwich compounds[46]. Relative to the cyclopentadienyl fragment, the χ^{MO} for Ni(NO) is in line with a strong donor like behaviour from the latter which is indeed corroborated experimentally[47]. As compared to the Ni fragments, the χ^{MO} for the carbonyl fragments of $Co(CO)_4$ and $Mn(CO)_5$ show enhanced donor property. A slightly smaller value of χ^{MO} for the latter as compared to the CH_3 group as well as the H atom suggests a possibility of the charge transfer from metal to CH_3 (or H) in the parent molecule formed by the metal carbonyl and CH_3 (or H); this is also detected experimentally[48]. The χ^{MO} values for the three $M(CO)_3$ fragments are intermediate between the typical acceptor (e.g. VO_4, CrO_4) and typical donor systems (NiNO, NiBr) indicating another well-known ability of such fragments to act both as donors and acceptors depending upon the other fragment bonding with it to form the parent molecule. It is interesting to compare χ^{MO} values for $Mn(CO)_3$ and Ni(NO) fragments as both are known to form complexes with the cyclopentadienyl group. Based on the calculated χ^{MO} values a relatively more predominant covalent bonding would be expected in the Mn-complex. This has been also corroborated by means of photoelectron studies on the two complexes by Lichtenberger et al.[49].

We shall conclude this section by comparing the various theoretical estimates of χ for a set of CX_3 fragments. Bergmann and Hinze[4] and Mortier[34] have concluded that the various semiempirical approaches based on PEE give a very good account of the atomic net charges in molecules. Böhm and Schmidt[24] have used a set of theoretically derived χ^{LDA} values for several atomic states and arrived at similar conclusions. A comparison of the various estimates of χ based on the PEE with χ^{MO} is made in Table 12. It is found that all data give similar trends. The numerical differences in χ^{MO} values corresponding to different CX_3 groups are less pronounced than those obtained from electronegativity equalization. This can be explained on the basis of an effective averaging of the various LCAO amplitudes in the active MO through the calculation of the one-electron eigenvalue which determines χ^{MO} (see also discussion relating to Table 8). Similarly, the divergent behaviour of χ^{MO} can be understood in terms of the breakdown of the one-

Table 12. Group electronegativities χ_{gp} (in eV) for some CX_3 radicals assuming sp^3-hybrid bonds of carbon, p bonds for halogen and s bonds for hydrogen[24)]

	Atom	q_a	χ_{gp}				
			BS[e]	BSS[a]	BH[b]	Mullay[c]	Huheey[d]
CH_3	C	−0.12	6.9	6.3	7.45	2.17	2.27
	H	0.04					
CHF_2	C	0.10	9.0	6.7	9.44		3.00
	H	0.22					
	F	−0.16					
CHFCl	C	0.05	8.5	6.5			2.82
	H	0.18					
	F	−0.20					
	Cl	−0.03					
$CHCl_2$	C	0.00	8.0	6.6	8.50		2.66
	H	0.14					
	Cl	−0.07					
CF_3	C	0.23	10.3			3.47	3.46
	F	−0.08					
CCl_3	C	0.06	8.6			2.89	2.84
	Cl	−0.02					

[a] M. C. Böhm, K. D. Sen, and P. C. Schmidt, data from the transition operator method, Ref. 22
[b] D. Bergmann and J. Hinze, Ref. 4
[c] J. Mullay, Ref. 3
[d] J. E. Huheey, Ref. 43
[e] M. C. Böhm and Schmidt, Ref. 24

electron picture as explained earlier in the case of the CH_2Cl fragment. We conclude that the theoretical studies reviewed in this chapter establish the approximate density functional theoretical approaches to the calculations of and as reliable tools in understanding chemical bonding and reactivity.

D. Future Scope of Work

The quantification of two of the most widely utilized concepts in chemistry, electronegativity and hardness, as realized in terms of the basic parameters in the density functional theory presents a significantly important development in quantum chemistry. The only criticism often leveled against the practical applications of calculating electronegativity and hardness is that the exact energy density functional is still unknown! The calculations reviewed in this chapter as well as those of others in this monograph establish that the use of simple local density approximations lead to reliable estimates of electronegativity and hardness parameters. The refinement of the calculated values is then directly related to the development in the basic theory. Successful attempts have been already made in this direction which lead to energy density functionals beyond the local approximation[50)]. Such efforts show definite indications of being able to realize the exact density functional in the near future. The advantages of the conceptual simplicity of such a theory in providing a broad-based understanding of the various structural phenomena in terms of electronegativity and hardness for atoms, molecules, clusters, and solids is yet another attractive feature for chemists. At the present state of the applications of the theory, it is clearly established that the calculations of the various properties are characterized by an inherently increased accuracy for the heavier atomic and molecular systems, and for the lighter systems they are quite reliable. The limitations of the theoretical approximations made in the actual calculations show up clearly trends in electronegativity and hardness values. The available energy density functionals should be now used to calculate these parameters for the atoms in various oxidation states, for example, M(I)–M(IV) in d-block elements and M(II)–M(IV) in f-block elements, respectively. Such calculations must be examined in terms of the basic task of rationalizing a large volume of empirical data on electronegativity[51)] available in the literature. Lastly, with the development of the geometry optimization techniques[52)] using the density-based approach, analysis of structure-reactivity data in terms of electronegativity and hardness also seems to be a practical possibility for the future.

Acknowledgement. The authors are grateful to Professor Alarich Weiss for his interest and constant encouragement throughout the development of this work. KDS thanks the Alexander von Humboldt Stiftung, F.R.G. and University Grants Commission, India for the grant of AvH fellowship and Career-Award respectively. PCS thanks the Fonds der Chemischen Industrie for financial support.

E. References

1. Pauling, L.: J. Am. Chem. Soc. *54,* 3570 (1932)
2. Mulliken, R. S.: J. Chem. Phys. *3,* 782 (1934)
3. Mullay, J.: This monograph, Chapt. 1
4. Bergmann, D., Hinze, J.: This monograph, Chapt. 7
5a. Iczkowski, R. P., Margrave, J. L.: J. Am. Chem. Soc. *83,* 3547 (1961)
5b. Pritchard, H. O., Sumner, F. H.: Proc. Roy. Soc. Ser. A 235, 136 (1956)
6a. Hohenberg, P., Kohn, W.: Phys. Rev. *136,* B 864 (1964)
6b. Kohn, W., Sham, L. J.: ibid. *140,* A 1133 1965)
7a. Rajagopal, A. I.: Adv. Chem. Phys. *41,* 59 (1980)
7b. Parr, R. G.: Ann. Rev. Phys. Chem. *34,* 631 (1983)
7c. Dahl, J. P., Avery, J. (Editors): Local Density Approximations in Quantum Chemistry and Solid State Physics; Plenum Press, New York – London 1984
8. Parr, R. G., Donnelly, R. A., Levy, M., Palke, W. E.: J. Chem. Phys. *68,* 3801 (1978)
9. Bartolotti, L. J.: This monograph, Chapt. 2
10. Alonso, J. A., Balbás, L. C.: This monograph, Chapt. 3
11. C. Froese Fischer: Atomic Hartree-Fock Theory, Wiley Interscience, New York 1977
12. Bagus, P. S.: Phys. Rev. *139,* A 619 (1965)
13. Slater, J.C.: The Self-Consistent Field Theory for Molecules and Solids, McGraw Hill, New-York 1974
14. Gunnarson, O., Lundquist, B. I.: Phys. Rev. *B 13,* 4274 (1976)
15. Vosko, S. H., Wilk, L., Nusair, M.: Can. J. Phys. *58,* 1200 (1980)
16. Goscinski, O., Pickup, B. T., Purvis, G.: Chem. Phys. Lett. *22,* 167 (1973);
Goscinski, O., Hehenberger, M., Roos, B., Siegbahn, P.: ibid. *33,* 427 (1975); Hehenberger, M.: ibid. *46,* 117 (1977);
Firsht, D., Pickup, B. T.: Chem. Phys. Lett. *56,* 295 (1978)

17. Pearson, R. G.: J. Am. Chem. Soc. *85,* 3533 (1963); Science *51,* 172 (1966)
18. Pearson, R. G.: Hard and Soft Acids and Bases, Dowden, Hutchinson and Ross, Stroudenburg, Pa (1973);
    Ho, T. L.: Hard and Soft Acid and Bases in Organic Chemistry, Academic Press, New York 1977;
    Jensen, W. B.: The Lewis Acid-Base Concept, Wiley-Interscience, New York 1980, Chapt. 8
19. Parr, R. G., Pearson, R. G.: J. Am. Chem. Soc. *105,* 7512 (1983)
20. Sen, K. D., Schmidt, P. C., Weiss, A.: Theoret. Chim. Acta *58,* 69 (1980)
21. Sen, K. D.: J. Phys. *B16,* L149 (1983)
22. Schmidt, P. C., Böhm, M. C.: Ber. Bunsenges. Phys. Chem. *87,* 925 (1983)
23. Sen, K. D., Schmidt, P. C., Böhm, M. C.: J. Phys. *B18,* 35 (1985)
24. Böhm, M. C., Schmidt, P. C.: Ber. Bunsenges. Phys. Chem. *90,* 913 (1986)
25. Böhm, M. C., Sen, K. D., Schmidt, P. C.: Chem. Phys. Lett. *78,* 357 (1981)
26. Böhm, M. C., Schmidt, P. C., Sen, K. D.: J. Mol. Str. (THEOCHEM) *87,* 43 (1982)
27. Sen, K. D., Böhm, M. C., Schmidt, P. C.: ibid. *106,* 271 (1984)
28. Sanderson, R. T.: Science *114,* 670 (1951)
29. Bratsch, S. G.: J. Chem. Ed. *61,* 588 (1984) for a general overview
30. Sen, K. D.: J. Phys. *B11,* L577 (1978); J. Chem. Phys. *71,* L1035 (1979); ibid. *75,* L1037 (1981); ibid. *75,* L5971 (1981)
31. Hadjisavvas, N., Theophilon, A.: Phys. Rev. *A32,* 720 (1985)
32. Böhm, M. C.: Z. Naturforsch. *36a,* 1205 (1981)
33. Böhm, M. C., Gleiter, R.: Theoret. Chim. Acta *59,* 127 (1981)
34. Gázquez, J. L., Vela, A., Galván, M.: This monograph, Chapt. 4
35. Mortier, W.: This monograph, Chapt. 6
36. Bartolotti, L. J., Gadre, S. R., Parr, R. G.: J. Am. Chem. Soc. *102,* 2945 (1980)
37. Liberman, D., Waber, J. T., Cromer, D. T.: Phys. Rev. *137,* A27 (1965)
38a. Cole, L. A., Perdew, J. P.: Phys. Rev. *A25,* 1265 (1982)
38b. Robles, J., Bartolotti, L. J.: J. Am. Chem. Soc. *102,* 2445 (1984) and also discussion at the end of Sect. IV A; Chapt. 2 of this monograph
39. Hotop, H., Lineberger, W. C..: J. Phys. Chem. Reference Data *4,* 439 (1973)
40. Schmidt, P. C., Böhm, M. C., Weiss, Al.: Ber. Bunsenges. Phys. Chem. *89,* 1330 (1985)
41. Shanker, S., Parr, R. G.: Proc. Natl. Acad. Sc. U.S.A. *82,* 264 (1985)
42. Hinze, J., Jaffe, H. H.: J. Am. Chem. Soc. *84,* 540 (1962)
43. Huheey, J. E.: J. Phys. Chem. *69,* 3284 (1965)
44. Pauling, L.: The Nature of Chemical Bond, Cornell Univ. Press, Ithaca, New York 1960
45. Schleyer, P. v. R., Tidor, B., Jemmis, E. D., Chandrasekhar, J., Würthwein, E.-U., Kos, A. J., Luke, B. T., Pople, J. A.: J. Am. Chem. Soc. *105,* 484 (1983)
46. Werner, H.: Angew. Chem. *89,* 1 (1977);
    Krüger, C., Sekutowski, J. C.: Angew. Chem. *89,* 179 (1977)
47. Cotton, F. A., Wilkinson, G.: Anorganische Chemie, Weinheim, Verlag Chemie 1970
48. McNeill, E. A., Scholer, F. R.: J. Am. Chem. Soc. *99,* 6243 (1977)
49. Lichtenberger, D. L., Fenske, R. F.: ibid. *98,* 50 (1976)
50. Langreth, D. C., Mehl, M. J.: Phys. Rev. *B28,* 1809 (1983)
51. Sanderson, R. T.: Inorg. Chem. *25,* 3518 (1986)
52. Dunlap, B. I.: J. Phys. Chem. *90,* 5524 (1986)

Electronegativity Equalization and its Applications

Wilfried J. Mortier

Department of Chemistry, University of North Carolina, Chapel Hill, North Carolina 27514, USA and K. U. Leuven, Laboratorium voor oppervlaktechemie, Kard. Mercierlaan 92, B-3030 Leuven (Heverlee), Belgium

Applications of the electronegativity equalization principle for qualitatively as well as quantitatively predicting intrinsic properties of atoms in a molecule (partial charges, charge reorganizations due to molecular interactions) and of reaction energies are reviewed and supplemented with original data. The emphasis is placed on the "effective" electronegativity of an atom in a molecule and its equalization, producing two chemically important parameters: the average electronegativity and partial charges. Several formalisms for obtaining this chemically pertinent information are compared. Depending on the property under investigation, these parameters may, separately or combined, characterize the physico-chemical properties of the molecule.

1. Introduction: Electronegativity Equalization . . . 126

2. Energy and Electronegativity of the Atom in a Molecule . . . 128
   a. Molecular Energy of Formation and Electronegativity: Empirical Relations . . . 128
      (i) Electrostatic and Covalent Energy . . . 128
      (ii) "Electronegativity Energy" and Interaction Energy . . . 129
   b. Corrections for the Valence State and the External Potential . . . 130
      (i) Valence State . . . 130
      (ii) External Potential and Covalent Contributions . . . 130
   c. Higher-Order Expansions . . . 131
   d. Separation of Variables . . . 132

3. Average Electronegativity and Partial Charges . . . 133
   a. Connectivity-Dependent Charges . . . 133
      (i) Total Electronegativity Equalization . . . 133
      (ii) Empirical Approximations . . . 134
   b. Methods Based on Isolated-Atom Electronegativities . . . 136
   c. Other Methods . . . 138

4. Parametrization . . . 139

5. Electronegativity Equalization and Chemical Information . . . 140

6. Prospects . . . 141

7. Note Added in Proof . . . 142

8. References . . . 142

1. Introduction: Electronegativity Equalization

Qualitatively, electronegativity is a readily comprehensible concept introduced by Pauling[1] as the power of an atom in a molecule to attract electrons to itself. The quantitative aspects were less clear for a long time, during which period several quantitative electronegativity scales were proposed, all of them defining to some extent the tendency of an atom in a molecule to attract electrons[2]. An important contribution to the quantitative expression of the electronegativity concept was made by Iczkowski and Margrave[2]. The intuitive basis for their definition of the electronegativity, i.e.

$$\chi = -\frac{dE}{dN} \tag{1}$$

is as follows. An energy lowering will result after joining two atoms if electrons are transferred to the atom for which the decrease in total energy dE by gaining a fractional electronic charge dN exceeds the increase in total energy of the atom from which these electrons are removed. An atom with a given slope of the total energy curve E(N) will take away electrons from any other atom which has a smaller slope, and in doing so, lower the systems energy. The electronegativity of an atom was therefore identified with this slope, and may be evaluated also for ions. For a variation of the energy according to a second-degree polynomial, and using the ionization potential I and the electron affinity A of an atom as the change in energy needed to remove or to add an electron respectively, Eq. (1) reduces for the neutral atom to

$$\chi^0 = \frac{(I + A)}{2} \tag{2}$$

i.e. Mulliken's[3] definition for the electronegativity.

Parr et al.[4,5] showed that this quantity is identical to the negative of the chemical potential μ of the electronic cloud. In density functional theory[6,7] μ appears as a Lagrange multiplier in the variational principle used for obtaining the ground state electron density ϱ (which is typical for a given external potential[8] v) as

$$\delta(E[\varrho, v] - \mu N[\varrho]) = 0 \quad \text{for constant v} \tag{3}$$

where $E[\varrho, v]$ is the energy as a functional of the electron density ($E[\varrho, v] = \int \varrho(\vec{r})v(\vec{r})d\vec{r} + F[\varrho]$; $F[\varrho]$ is the functional containing the kinetic energy and the electron-electron repulsion terms) and $N[\varrho] = \int \varrho d\tau$, i.e. the total number of electrons. For the true density ϱ satisfying Eq. (3), since $\delta N/\delta\varrho = 1$,

$$\mu = \frac{\delta E}{\delta\varrho(\vec{r})} = v(\vec{r}) + \frac{\delta F[\varrho]}{\delta\varrho(\vec{r})} \tag{4}$$

has the same value everywhere. Also, since $E = E[N, v]$,

$$dE = \left(\frac{\partial E}{\partial N}\right)_v dN + \int \left(\frac{\delta E}{\delta v(\vec{r})}\right)_N dv(\vec{r})\, d\vec{r} \qquad (5)$$

such that

$$\mu = \left(\frac{\partial E}{\partial N}\right)_v \qquad (6)$$

and

$$\varrho(\vec{r}) = \left(\frac{\delta E}{\delta v(\vec{r})}\right)_N \qquad (7)$$

Equation (6) identifies μ with $-\chi$ [Eq. (1)], and since it is constant throughout the molecule, Sanderson's[9, 10] postulate that when two or more different atoms combine to form a molecule, their electronegativities change to a common intermediate value, is hereby validated. The analogy with macroscopic thermodynamics in defining μ as the chemical potential [Eq. (6)] is clear.

Independent of any particular theoretical framework, Politzer and Weinstein[11] proved for molecules in the ground state, that the electronegativities of all arbitrary portions of the total number of electrons are the same. This follows from the following considerations for diatomic molecules, but the approach can easily be extended to polyatomic molecules. The total energy is a function of the number of electrons associated with each atom (N_α, N_β), the nuclear charges, (Z_α, Z_β) and the internuclear distance R. For constant Z_α and Z_β it is possible to write

$$dE = \left(\frac{\partial E}{\partial N_\alpha}\right)_{R, N_\beta} dN_\alpha + \left(\frac{\partial E}{\partial N_\beta}\right)_{R, N_\alpha} dN_\beta + \left(\frac{\partial E}{\partial R}\right)_{N_\alpha, N_\beta} dR \qquad (8)$$

In the ground state and at equilibrium ($R = R_e$ and $dE = 0$ for any infinitesimal electron transfer $dN = -dN_\alpha = dN_\beta$) Eq. (8) reduces to

$$\left(\frac{\partial E}{\partial N_\alpha}\right)_{R_e, N_\beta} = \left(\frac{\partial E}{\partial N_\beta}\right)_{R_e, N_\alpha} \qquad (9)$$

stating that the electronegativities of the two atoms in the molecule are equal in the ground state.

The molecular energy E in Eq. (9), and therefore also the "effective" atomic electronegativities will depend on the equilibrium charge distribution as well as on the molecular geometry. If now the explicit dependence of the individual atomic electronegativities on the charge transfer in the molecule and on the geometry are known, the principle of electronegativity equalization is all that is required to calculate the atomic charges. This explicit dependence is the subject of the following section.

2. Energy and Electronegativity of the Atom in a Molecule

The electronegativity at any point in an atom or in a molecule could be evaluated by applying Eq. (4). The external potential at any point $v(\vec{r})$ is the nuclear-electron potential $-Z/r$, which for a molecule includes the interaction with all nuclei. The explicit expression for $F[\varrho]$ is unknown, and is a topic of research. In order to maintain the idea of an atom in a molecule, it is required to partition the electron density ϱ in atomic regions[12]. Even in case appropriate partitionings of the electron density can be made, it is obvious from Eq. (4) that the electronegativity of an atom in a molecule cannot be identical to its isolated-atom value: charge transfer might have occured, the external potential has changed and the kinetic energy and the electron-electron repulsion energy, both included in $F[\varrho]$, will differ, not only because of a different number of electrons, but also because of a change in the size and shape of the electronic cloud. Nevertheless, before any applications can be thought of, we need an explicit expression for the electronegativity of an atom in a molecule.

Fortunately, there are other ways to proceed. If indeed electrons can be assigned to certain regions in space, such as N_α electrons belonging to atom α in the molecule, and we dispose of an expression for the total energy in terms of N_α, N_β ... explicitly, we may apply Eq. (9) to calculate the chemical potential of this region. Several authors have been able to successfully calculate molecular (mostly diatomic) energies of formation, using empirical data and atomic electronegativities as input. Their expressions will now be investigated for their potential use for explicitly deriving the electronegativity of atoms in a molecular environment.

Before we proceed, some definitions are required as to how the electronegativity χ or the chemical potential μ $(= -\chi)$ can be obtained from an empirical expression for the total energy function E(N) by applying Eq. (1). For an isolated atom, E(N) is most easily approximated by a polynomial[13)]

$$E_\alpha = E_\alpha^0 + \mu_\alpha^0 (N_\alpha - N_\alpha^0) + \eta_\alpha^0 (N_\alpha - N_\alpha^0)^2 + \dots \tag{10}$$

with μ_α^0 the chemical potential of the neutral atom α $(N_\alpha = N_\alpha^0)$, i.e. dE_α/dN_α, and η_α^0 defined by Parr and Pearson[13)] as the absolute hardness, i.e. $1/2(d^2E_\alpha/dN_\alpha^2)$. The variation of the isolated-atom electronegativity as a function of the charge q_α $(\Delta N_\alpha = N_\alpha - N_\alpha^0 = -q_\alpha)$ is then given by

$$\chi_\alpha = \chi_\alpha^0 + 2\eta_\alpha^0 q_\alpha \tag{11}$$

For the calculation of the electronegativity, a knowledge of the binding energy is sufficient since the total energy of the isolated neutral atom, E_α^0 in Eq. (10), disappears when taking the derivative.

a. *Molecular Energy of Formation and Electronegativity: Empirical Relations*

(i) The single most successful approach was followed by Sanderson[9, 10, 14)]. Energies of hundreds of compounds were calculated by evaluating the binding energy in each bond separately as the sum of an electrostatic and a covalent contribution, weighted by an ionic weighting coefficient $t_i = (q_\alpha - q_\beta)/2$. The charges are obtained from the equalization of the atomic electronegativities with the geometric average compound electronegativity. For diatomic molecules $t_i = q = (\chi_\beta^0 - \chi_\alpha^0)/2(\eta_\alpha^0 + \eta_\beta^0)$, and the energy of formation is given by

$$E = -t_i\frac{e^2}{R_{\alpha\beta}} - (1 - t_i)\frac{(r_\alpha + r_\beta)}{R_{\alpha\beta}}(E_{\alpha\alpha}E_{\beta\beta})^{1/2} \tag{12}$$

with r_α, r_β the nonpolar covalent radii, and $E_{\alpha\alpha}, E_{\beta\beta}$ the homonuclear single covalent bond energies.

A similar approach was made by Matcha[15], by proposing a relation between bond energies and electronegativity differences viz.

$$E = 252\left(\frac{f}{R_{\alpha\beta}}\right) + (1 - f)\frac{(E_{\alpha\alpha} + E_{\beta\beta})}{2} \tag{13}$$

where f is the effective charge transferred (approximately related to the partial charges by $f = (q_\alpha - q_\beta)/2$, and estimated using an empirical relationship $f/R_{\alpha\beta} \approx 2/3\,(1 - \exp[(\chi_\alpha - \chi_\beta)^2/4]))$.

Sanderson's and Matcha's approach have in common that only ionic and covalent contributions to the binding energy are considered, weighted by an ionicity derived from the electronegativity differences, but with no explicit reference to the electronegativities of the atom in a molecule. Expressions (12) and (13) cannot be used for deriving the electronegativity of an atom in a molecule by applying Eq. (9): a separation of the variables in $N_\alpha, N_\beta \ldots$ is not possible.

(ii) Evans and Huheey[16] analyzed the binding energy in terms of "electronegativity energy" [i.e. the energy expressed by Eq. (10)], Madelung energy, and covalent energy. For a diatomic molecule, the following expression can be written (using $q_\alpha = -q_\beta = q$):

$$E = \chi_\alpha^0 q + \eta_\alpha^0 q^2 - \chi_\beta^0 q + \eta_\beta^0 q^2 - \frac{q^2}{R_{\alpha\beta}} A\left(1 - \frac{1}{n}\right) - (1 - q^2)E_H \tag{14}$$

with A the Madelung constant, n the Born exponent and E_H the "homopolar energy" of a heteronuclear bond (for definitions, see Ref. 16). The actual charge q is determined by finding the minimum of the total energy function, which occurs for $dE/dq = 0$:

$$\frac{dE}{dq} = 0 = \chi_\alpha^0 + 2\eta_\alpha^0 q - \chi_\beta^0 + 2\eta_\beta^0 q - 2\frac{q}{R_{\alpha\beta}} A\left(1 - \frac{1}{n}\right) + 2qE_H \tag{15}$$

This is exactly equivalent to an electronegativity equalization. Because of the fact that E_H contains the product of the homopolar energy of both atoms however, it's still not possible to factorize Eq. (15) in terms of the contributions of the atoms α and β separately. It is nevertheless instructive to see that other contributions than just the change of the electronegativity with charge [Eq. (11)] will appear in an expression for the electronegativity of an atom in a molecule.

Ferreira[17, 18] considers the bond stabilization energy as this part of the binding energy that exceeds the covalent bond energy [i.e. $(E_{\alpha\alpha} \cdot E_{\beta\beta})^{1/2}$]. The bond stabilization energy can then be separated in the electrostatic energy $[q^2(1 - 1/n)/R_{\alpha\beta}]$ and a contribution from the partial electron affinity of the more electronegative atom $A_\alpha(2q - q^2)$ and the partial ionization energy of the less electronegative atom $-I_\beta q^2$. The charges q are estimated from the differences in electronegativity and the sum of the hardnesses as $q = (\chi_\alpha^0 - \chi_\beta^0)/2(\eta_\alpha^0 + \eta_\beta^0)$ where a distinction is made in the hardnesses, depending on the removal or acceptance of the electron respectively. This expression is different from Eq. (15), in that part of the electronegativity energy (χ_α^0 may be replaced by Eq. (2) and η_α^0 by $(I - A)/2$, see below) will be absorbed in the unweighted contribution of the covalent binding energy to the total energy. Again, there is no separation of the variables, which could allow the application of Eq. (9).

b. Corrections for the Valence State and the External Potential

(i) Ponec[19] generalized the orbital electronegativity concepts introduced by Hinze et al.[20] to calculate global atomic electronegativities in the CNDO approximation. The total molecular energy was partitioned in mono- and biatomic components, using an appropriate partitioning scheme. It was postulated that only monoatomic terms contribute to the "energy of an atom in a molecule", such that overlap contributions and Madelung terms are neglected in the definition of the electronegativity of an atom in a molecule. Within these limitations, it was stressed that the electronegativity is not invariant, but that changes in the valence state have to be considered in going from one molecule to another, viz.

$$\chi_\alpha = \frac{\sum_{i}^{\alpha} p_{ii}\chi_i^\alpha}{\sum_{i}^{\alpha} p_{ii}} \tag{16}$$

where the sum over all orbital electronegativities of atom α (χ_i^α) is weighted by the individual charge densities p_{ii}. The individual orbital electronegativities were then defined as

$$\chi_i^\alpha = -U_{ii}^\alpha - \left(P_\alpha - \frac{1}{2}\right)\gamma_{\alpha\alpha} \tag{17}$$

with $\gamma_{\alpha\alpha}$, the repulsion integral, and U_{ii}^α as atomic parameters, and where the atomic charge density P_α is the only variable.

(ii) Reed[21] emphasizes that although electronegativity equalization is rigorously true, it is only approximately so if isolated-atom electronegativities are equalized since this fails to minimize the system's energy if electrostatic interactions (H^{ES}) and resonance energies are not considered. The total number of electrons should therefore obey the following relation (which is again equivalent with the equalization of the atomic electronegativities):

$$\frac{dE}{dN_n} = 0 = \sum_{\alpha} (\chi_\alpha^0 + 2\eta_\alpha^0 N_\alpha) \frac{dN_\alpha}{dN_n} + \frac{d}{dN_n} H^{ES} + \frac{d}{dN_n} \left(\sum_{m \neq \alpha} \sum q_{m\alpha} H_{m\alpha} \right) \tag{18}$$

where $q_{m\alpha}$ is the $m\alpha$ bond order and $H_{m\alpha}$ the resonance energy.

Nethercot[22] clearly states that in an expansion of the electronegativity [such as Eq. (11)], the constants may differ from the free-atom values from compound to compound due to interatomic interactions. He also shows that the Madelung correction term can be incorporated in these constants. Incorporating covalent effects and electrostatic effects into the constants of expansion however is equivalent with requiring different coefficients for every atom in the molecule.

For an overall energy expression of a molecule, accepting a parabolic expression for the isolated-atom energies [equivalent with Eq. (10)], and containing an ionic and a covalent contribution as

$$E = \sum_{\alpha} E_\alpha^0(q_\alpha) + e^{ion}(q_\alpha, q_\beta \ldots) + e^{cov} \tag{19a}$$

Komorowski[23] corrects the chemical potential of the αth atom as

$$\mu_\alpha = \mu_\alpha^0(q_\alpha) - V_\alpha \quad \text{where} \quad V_\alpha = \frac{\partial e^{ion}}{\partial q_\alpha} \tag{19b}$$

assuming e^{cov} not to depend on the partial charges. A similar expression was used by Balbas et al.[24] (for diatomic molecules where V_α can be explicitly calculated as $V_\alpha = q_\alpha/R_{\alpha\beta} = -q_\beta/R_{\alpha\beta}$). This obviously includes only a correction to $v(\vec{r})$ in Eq. (4), but not to $\delta F[\varrho]/\delta\varrho$.

c. *Higher-Order Expansions*

A major shortcoming of an expansion of the energy of the atom in the molecule as a function of the number of electrons only [Eq. (10)] is that it neglects perturbations due to the external potential. A full second-order expansion was extensively investigated recently by Nalewajski[25–27] for an energy function $E = E(N, Z)$:

$$\begin{aligned} E(N,Z) &= E^0 + \left\{ \frac{\partial E}{\partial N} \Delta N + \frac{\partial E}{\partial Z} \Delta Z \right\} \\ &\quad + \frac{1}{2} \left\{ \frac{\partial^2 E}{\partial N^2} (\Delta N)^2 + 2 \frac{\partial^2 E}{\partial N \partial Z} \Delta N \Delta Z + \frac{\partial^2 E}{\partial Z^2} (\Delta Z)^2 \right\} \\ &= E^0 + \{\mu^0 \Delta N + v^0 \Delta Z\} + \{\eta^0 (\Delta N)^2 + 2\alpha \Delta N \Delta Z + \beta (\Delta Z)^2\} \end{aligned} \tag{20}$$

giving for the chemical potential of the atom in a molecule

$$\mu(N,Z) = \mu^0 + 2(\eta^0 \Delta N + \alpha \Delta Z) \tag{21}$$

where $\alpha = 1/2(\partial\mu/\partial Z)_N = 1/2(\partial v/\partial N)_Z$. ΔZ, the increase in core charge, accounts for the fact that in the molecule, the outer electrons are in the presence of all atomic cores. The magnitude of ΔZ will depend on the nature of its partners. An explicit dependence of $\Delta Z_\alpha(\beta, R)$ (for atom α in a diatomic molecule depending on its bonding partner β and the interatomic distance R) was proposed by Nalewajski[26, 27]. This author also stresses the importance of including the external potential (ΔZ) in the equalization of the chemical potential. The terms μ^0 and $\alpha\Delta Z$ can be combined in μ^*, i.e. a valence-state potential which includes the change in external potential.

d. Separation of Variables

Although there is no agreement among the various authors about an expression for the electronegativity of an atom in a molecule, there is a general understanding that it differs from the free-atom value due to valence-state effects, covalent effects as well as electrostatic effects occurring in chemical bonding. Faced with several possible partitionings of the chemical potential and of the total energy, a general applicability requires that no pertinent information is lost, and that all quantities are easily calculated. The external potential, for example, contains information on the connectivity, and should not be integrated in any other term. Such an empirical formalism was proposed by Mortier et al.[28]. A theoretical basis for it was given by Mortier, Ghosh and Shankar[29].

In order to maintain the concept of an atom-in-a-molecule, the total molecular energy can be partitioned in a sum of atomic contributions:

$$E(N_\alpha, N_\beta \ldots;\; Z_\alpha, Z_\beta \ldots;\; R_{\alpha\beta} \ldots) = \sum_\alpha \langle \hat{T}^\alpha + \hat{V}^\alpha_{ne} + \hat{V}^\alpha_{ee} + \hat{V}^\alpha_{nn} \rangle \tag{22}$$

i.e. the sum of the expectation values of the kinetic energy, the nuclear-electron (ne) and electron-electron (ee) interaction energies, and the nuclear-nuclear (nn) repulsion energies. By a factorization into typical atomic terms and atom-atom interaction terms, and assuming spherically symmetric "atomic" charge clouds, the total energy can be approximated as

$$\begin{aligned} E = & \sum_\alpha \left[T^{\alpha,\,mol} + V^{\alpha,\,mol}_{ne} + V^{\alpha,\,mol}_{ee}\right] \\ & + \sum_\alpha \left[-N_\alpha \sum_{\beta \neq \alpha} \frac{Z_\beta}{R_{\alpha\beta}} + \frac{1}{2} \sum_{\beta \neq \alpha} \frac{N_\alpha N_\beta}{R_{\alpha\beta}}\right] + \frac{1}{2} \sum_\alpha \sum_{\beta \neq \alpha} \frac{Z_\alpha Z_\beta}{R_{\alpha\beta}} \end{aligned} \tag{23}$$

The "atomic terms", i.e. the first sum, includes charge-transfer effects (every term depends on N_α), and is different from the same terms for the isolated atom. It was proposed[29] that these terms be expanded similarly as for the free atom [Eq. (10)], but with μ^*_α and η^*_α as coefficients. For $\mu^*_\alpha = \mu^0_\alpha + \Delta\mu_\alpha$ and $\eta^*_\alpha = \eta^0_\alpha + \Delta\eta_\alpha$, the chemical potential of each atom is obtained using Eq. (9). The expression of the effective electronegativity of the atom in the molecule is then:

$$\chi_\alpha = (\chi^0_\alpha + \Delta\chi_\alpha) + 2(\eta^0_\alpha + \Delta\eta_\alpha)q_\alpha + \sum_{\beta \neq \alpha} q_\beta / R_{\alpha\beta} \tag{24}$$

3. Average Electronegativity and Partial Charges

The equalization of the electronegativities of all atoms in a molecule represents a set of simultaneous equations (for n atoms, there are n − 1 equations of the type $\chi_\alpha = \chi_\beta$), subject to the condition of the conservation of charges ($\sum_\alpha q_\alpha$ = constant). The average molecular electronegativity $\bar{\chi}$ (= $\chi_\alpha = \chi_\beta \ldots$) and the charges of the atoms in the molecule, both important quantities, can be calculated directly by solving this set of equations.

If the atomic electronegativities are represented by equations of the type (11), no connectivity information is included and for all atoms of the same type in a molecule (i.e. with the same χ^0 and η^0), identical charges will be calculated. This is contrary to chemical intuition although, as will be shown at a later stage, valuable information can still be derived from these simplified methods. For an electronegativity expression such as Eq. (24), the explicit incorporation of the Madelung term into the electronegativity ensures that all information about the molecular conformation is retained, and no two atoms will be identical, unless they are placed in exactly the same environments. This method will be discussed first and will serve as a basis for comparison.

a. Connectivity-Dependent Charges

(i) Total Electronegativity Equalization

The application of Eq. (24) requires (i) an electronegativity scale as a function of the charges (χ^0 and η^0, see Sect. 4), (ii) the molecular configuration ($R_{\alpha\beta}$) and (iii) a calibration of $\Delta\chi$ and $\Delta\eta$. It was demonstrated by Mortier, Ghosh and Shankar[29] that $\Delta\chi$ and $\Delta\eta$ can be calibrated so as to reproduce ab-initio charges, and that these can be transferred to different molecular configurations. $\Delta\chi$ and $\Delta\eta$ will obviously depend on the charges used for calibration and on the electronegativity scale. In a way, the choice of the electronegativity scale is not so important, since any "errors" in χ^0 and η^0 are incorporated in $\Delta\chi$ and $\Delta\eta$ anyway. The transferability of $\Delta\eta$ and $\Delta\chi$ to different molecular configurations is remarkable. Since $\Delta\chi$ and $\Delta\eta$ represent changes in the isolated-atom electronegativity and hardness due to the confinement of the atom in a molecule, this transferability is not unrestricted. The charge transfers in conjugated systems, for example, are not reproduced properly and probably require a more explicit expression for the covalent and resonance effects, now absorbed in $\Delta\chi$ and $\Delta\eta$. This formalism was called FEOE[28] (Full Equalization of Orbital Electronegativity), to distinguish it from the PEOE formalism[28, 30] (Partial Equalization of Orbital Electronegativity), which was developed first, and for which it constitutes the theoretical basis (see below).

The applicability of the FEOE formalism is illustrated for a series of molecules (CH_4, CH_3F, CH_2F_2, CHF_3, CF_4, ethane, propane and 1-F ethane; using Sanderson's[14] scale; $\Delta\chi$ and $\Delta\eta$ for C, H and F calibrated to 1-F ethane, methane and CF_4) in Fig. 1, for which the geometries and the ab-initio charges were taken from the work by Hehre and Pople[31]. Slightly better correlations were obtained before for the F-substituted methane series[29], which is probably due to a larger and more realistic variation of the charges on F, as calculated by Brundle et al.[32]. This illustrates the importance of the choice of the set of charges for $\Delta\chi$ and $\Delta\eta$.

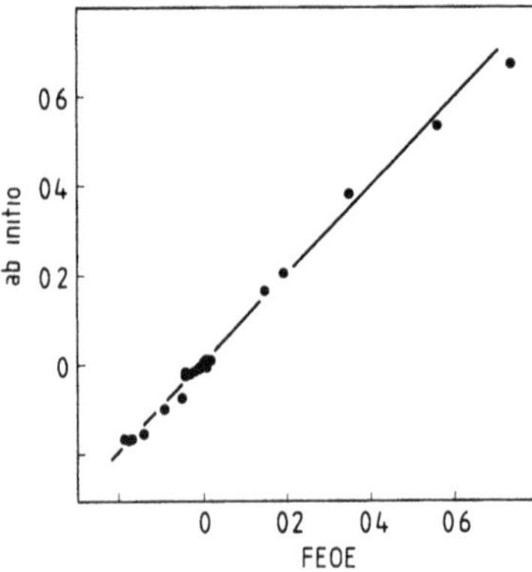

Fig. 1. Correlation between ab-initio charges and charges calculated with the FEOE method (after calibration of $\Delta\chi$ and $\Delta\eta$; Sanderson scale) for C, H, and F in CH_4, CH_3F, CH_2F_2, CHF_3, CH_3–CH_3, $CH_3CH_2CH_3$ and CH_3CH_2F

An analysis of the effective electronegativity [(Eq. (24)] is interesting, and forms the basis of several successful approximations allowing only a partial equalization of the atomic electronegativities. For an atom surrounded by positive charges ($\Sigma q_\beta/R_{\alpha\beta}$ positive), an increase in the effective electronegativity is predicted. The reverse is true for an atom surrounded by negative charges. This is exactly equivalent with Gutmann's[33] pileup and spillover of electrons at the donor site (negatively charged; electronegativity increases due to the presence of a positively charged acceptor) and at the acceptor site respectively in molecular interactions. The importance of the Madelung terms was further illustrated in Ref. 29, where charge shifts due to interactions between water molecules in different configurations[35] were very accurately reproduced. The presence of another molecule in the vicinity, even without charge transfer between them, will affect the Madelung term, and the charges will be adjusted throughout the molecule. In the case of molecular interactions, changes in $\Delta\chi$ and $\Delta\eta$ will be negligible with respect to changes in the electrostatic interactions. The FEOE method, as originally presented[28] did not contain a constant term $\Delta\chi$, while $2\,\Delta\eta$ was estimated to be proportional (s) to 1/r: s_α/r_α. The constant term $\Delta\chi$ is e.g. necessary to explain charge separations in homonuclear compounds (such as metal clusters) where not all atoms have the same environments. If there is no interest in calculating "absolute" charges, $\Delta\chi$ and $\Delta\eta$ need not be calibrated. Excellent shifts with C_{1s} binding energies (ESCA shifts) were obtained by putting $\Delta\chi = 0$ and $2\,\Delta\eta = 1/r_\alpha$, i.e. the former definition of the FEOE formalism[28].

(ii) Empiricial Approximations

The PEOE formalism[30] has considerable advantage and gives results that are very highly correlated with the FEOE formalism[28]. The PEOE formalism is an iterative method, evaluating the charge shifts in each bond separately. Only first neighbours are considered in each cycle with charge shifts dq calculated in a single bond as:

$$dq = 0.5^n(\chi_\alpha - \chi_\beta)/\chi_\beta^+ \tag{25}$$

where χ^+ is the electronegativity of the positive ion, and n the cycle number. The atomic charges, and therefore also χ_α, χ_β ... are adjusted in each cycle. Only the connectivities are required (no interatomic distances) as supplementary input to the electronegativity scale. The electronegativities are only partially equalized. This is realized by a smaller electron transfer than a full equalization would require [(Eq. (25)]. Isolated-atom electronegativities of negatively charged atoms are therefore kept at a higher, and those of positively charged atoms at a lower level, and in this way simulates the FEOE model. Extensions to conjugated systems were considered[36)] by treating σ and π electronegativities separately. The charges obtained for the σ-system are corrected by charge transfers in the π-system in accordance with the different resonance forms (forcing the π-charges to alternate).

Connectivity-information was incorporated in the process of electronegativity equalization by Jolly and Perry[37)] by considering the electronegativity equalization in each bond separately, an explicit contribution of the orbitals to χ^0 and a factorization of the charge transfer q_α in Eq. (11) as:

$$\chi_{nm} = \chi(p)_n + \frac{S_{nm}}{(N_{nm})^a}[\chi(s)_n - \chi(p)_n] + h_n\left[\frac{bq_{nm}}{(N_{nm})^a} + \sum_{l \neq m} q_{nl} + cF_n\right] \tag{26}$$

χ_{nm} represents the electronegativity of the orbital of an atom n used to form a bond to another atom m. The first two terms represent a weighted average of the electronegativities of the orbitals involved: s and p, with S_{nm} the fractional s character of the σ orbital and N_{nm} the bond order; h_n is proportional to $2\eta^0$; q_{nm}, is the negative charge transferred from atom n to m (separating the charge transferred in the bond from all other charges transferred to the same atom); F_n is the "formal" charge of atom n and a, b and c adjustable parameters, which were calibrated so as to correlate with experimental core-electron binding energies. Good results were obtained as compared to other methods, as well as for predicting experimental data. A computer program CHELEQ for making these atomic charge calculations was developed.

While Sanderson[9, 10)] postulated the average electronegativity to be the geometric average of the isolated-atom electronegativities (see below) and as such does not contain any connectivity information, the modified Sanderson method[38, 39)] does account for different electronegativities for atoms in different environments. The electronegativity of an atom in a molecule is calculated from the geometric average of the isolated-atom electronegativity and the average electronegativities of all groups attached to it ($\bar{\chi}_g$) as

$$\chi_\alpha = \left(\chi_\alpha^0 \prod_1^n \bar{\chi}_g\right)^{1/(n+1)} \qquad \text{with} \quad \bar{\chi}_g = \left(\prod_\beta^m \chi_\beta^0\right)^{1/m} \tag{27}$$

for n groups and m atoms in a group. Again, good correlations with experimental data were obtained.

A comparison of the modified Sanderson method, Jolly's method, CNDO and MINDO MO calculation was made recently by Linert et al.[40)] in a study of the charge distributions in alkylboranes and alkylamines, and their adducts with NH_3 and BH_3 respectively. Although all methods are in qualitative agreement for the isolated compounds, only the modified Sanderson method reproduced the pileup and spillover effects for the interacting molecules (as ab-initio as well as the FEOE method predict). Pileup and spillover effects are also not reproduced by the PEOE method[48)]. Comparing the modified Sanderson method with the FEOE method however, only the FEOE method

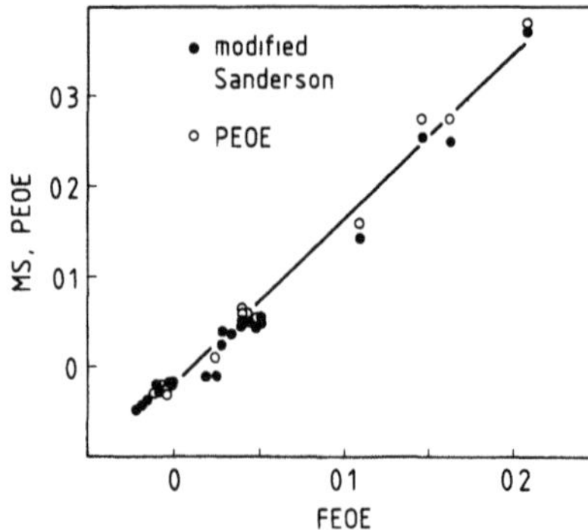

Fig. 2. Correlation between C-charges in several organic molecules (Ref. 28; Table 1) calculated with the FEOE method and charges calculated using the FEOE and Modified Sanderson formalisms (Sanderson scale used throughout)

will be able to account for different relative positionings of the molecules, because of the explicit use of interatomic distances.

In order to account for a correct prediction of the group electronegativity, depending on the site of interaction, Huheey[42] also considered a partial equalization (80%). This is best illustrated for a diatomic group AB–. The electronegativity of the substituted group is the electronegativity of B as perturbed by atom A. For uncharged groups, $\chi_{AB} = \chi_B$ can be estimated as

$$\chi_B = \chi_{AB} = \frac{2\eta_B^0\{(1-p)\chi_B^0 + p\chi_A^0\} + 2\eta_A^0\{\chi_B^0\}}{2(\eta_A^0 + \eta_B^0)} \tag{28}$$

with $p = 1.00$ for complete equalization, for which it is then identical with the average molecular electronegativity for the equalization of isolated-atom electronegativities [(Eq. (30)]. For $\chi_B^0 > \chi_A^0$, it means that χ_B is larger than in the case of a full equalization, which is equivalent to the PEOE method.

The two most easily applicable methods for simulating the full equalization of the effective electronegativities (the FEOE method) are the PEOE method and the modified Sanderson method. Both produce excellent correlations with charges obtained by FEOE, as illustrated in Fig. 2 for the same molecules considered in Ref. 28 (Sanderson's scale was used for both). A calibration to reproduce "absolute" charges is only possible in the FEOE formalism.

b. Methods Based on Isolated-Atom Electronegativities

The majority of the calculations using electronegativity equalization as a basis were performed without corrections to the isolated-atom electronegativities. The question to be asked is how these methods are related to an equalization of the "effective" electronegativities as defined in Eq. (24). It is obvious at least, that no connectivity-dependent electronegativities can be obtained in this way and that partial charges do not carry any supplementary information.

Sanderson[9, 10] postulated a geometric average electronegativity. For the same molecules as used for the calculation of the partial charges in Fig. 1, the geometric average electronegativity is plotted in Fig. 3 against the average electronegativity obtained by the FEOE formalism. The results are highly correlated for the isomorphous series of F-substituted methanes, but deviate slightly for CH_3CH_2F. This is an important limitation of its applicability, but at the same time reassuring that the geometric average electronegativity does in fact contain important information on the molecular properties for homologous series of compounds.

The validity of the geometric average of the isolated-atom electronegativities as being representative for the average molecular electronegativity was discussed by several authors. In the case where the valence-state atomic energies decay exponentially with the number of electrons, with the decay parameter being the same for all bonded atoms, it was shown by Parr and Bartolotti[43] that the geometric mean principle for electronegativity equalization is valid. This was also found to be only approximately true. Nalewajski et al.[26], analyzing the geometric average calculated from Eq. (21), conclude that the extra terms never vanish, but that it is small for soft atoms and that large deviations are expected for hard atoms. Nalewajski[27] further proposes the use of the valence-state chemical potential rather than the isolated-atom chemical potential μ^0. Parr and Bartolotti[43] also emphasize that the geometric mean principle does not trivially extend to a prediction of molecular electronegativities from functional groups. It has indeed already been discussed that this constitutes the basis of the modified Sanderson method [(Eq. (27)] which produces connectivity-dependent electronegativities and not the average molecular electronegativity.

For diatomic molecules, an equalization of the atomic electronegativities [(Eq. (11)] allows the calculation of the partial charges and the average electronegativity as

$$q = q_\alpha = -q_\beta = \frac{\chi_\beta^0 - \chi_\alpha^0}{2(\eta_\alpha^0 + \eta_\beta^0)} \tag{29}$$

$$\bar{\chi}_{\alpha\beta} = \frac{2\eta_\alpha^0\chi_\beta^0 + 2\eta_\beta^0\chi_\alpha^0}{2(\eta_\beta^0 + \eta_\alpha^0)} \tag{30}$$

Ferreira[17, 18] uses different values for the hardnesses of the cations and the anions. This is not unreasonable, in view of possible discontinuities in the E(N) vs. N curve[44].

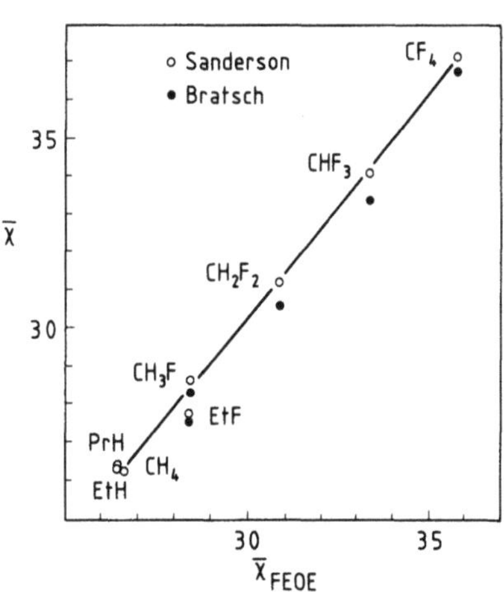

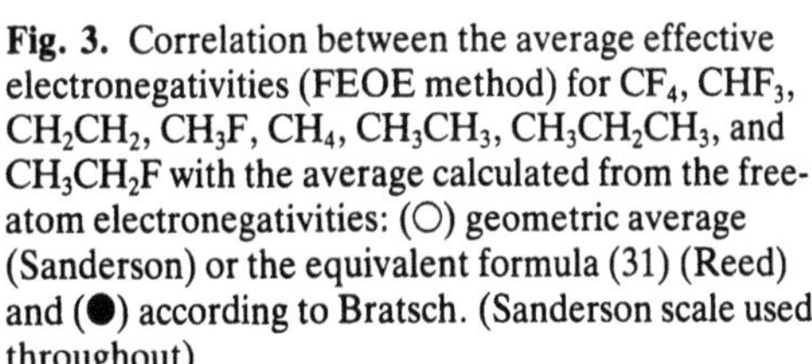

Fig. 3. Correlation between the average effective electronegativities (FEOE method) for CF_4, CHF_3, CH_2CH_2, CH_3F, CH_4, CH_3CH_3, $CH_3CH_2CH_3$, and CH_3CH_2F with the average calculated from the free-atom electronegativities: (○) geometric average (Sanderson) or the equivalent formula (31) (Reed) and (●) according to Bratsch. (Sanderson scale used throughout)

Huheey[45] also derived expressions for selected molecular compositions. Equation (29) was used by several authors (see e.g. Refs. 13, 17, 18, 24, 46). An analysis of Eq. (29) readily shows that the charge transfer is not uniquely determined by the electronegativity differences, but that it is attenuated by the sum of hardnesses[13]. Ferreira[17] also stressed this point and emphasises the fact that the ionicity is a property of each type of molecule, and that a single curve connecting the ionic character with the electronegativity difference has no meaning.

For polyatomic molecules, a convenient way to calculate the average molecular electronegativity (and therefore also by using Eq. (11) the partial charges) was derived by Reed[21]:

$$\bar{\chi}_{molec} = a_{molec} + \frac{1}{n} b_{molec} Z \qquad (31)$$

with $b_{molec.} = (\overline{1/2\eta^0})^{-1}$ and $a_{molec.} = (\overline{\chi^0/2\eta^0}) b_{molec.}$

where n is the number of atoms and $Z = \sum_{1}^{n} q_i$. Applying Eq. (31) for calculating the average molecular electronegativities for the series of molecules in the Figs. 1 and 3 yields exactly the same number as the geometric average electronegativity. This is due to the fact that in Sanderson's scale, $2\eta^0 \equiv k(\chi^0)^{1/2}$. A more simplified formula given by Bratsch[47, 48] does not reproduce the same numerical values ($\bar{\chi}_{molec.} = (n + Z)/\sum_{\alpha} (1/\chi_{\alpha}^0)$) (see also Fig. 3).

c. *Other Methods*

Komorowski[49] postulated that the chemical potential of the electrons in any atom or chemically stable neutral molecule is equal to zero. This forces the E(N) curve to have no linear or quadratic terms in N.

Inamoto and Masuda[50] presented a method for calculating group electronegativities based on Gordy's[51] equation for the electronegativity (energy/electron). Reasonable values of the group electronegativities of more than 150 substituents were obtained. For details of the method, see Ref. 50.

An equation for calculating orbital electronegativities for any atom or group was also proposed by Mullay[52]. The detailed expression depends on Slater's effective principal quantum number and modified screened nuclear charge, the percentage of p character of the orbital and the charge transferred from the surrounding bond forming orbitals.

4. Parametrization

An appropriate electronegativity scale, giving the electronegativity as a function of the charge, is required to calculate reasonable atomic charges. A detailed analysis of the different electronegativity scales is beyond the scope of this review, but its choice has consequences for the numerical results that are obtained from the different formalisms.

The E(N) curve can be constructed from experimental data. The use of the ionization potential and the electron affinity was already given in the introduction. For a simple quadratic variation, the expression for the electronegativity is then equivalent with Mulliken's definition [Eq. (2)] and the hardness is obtained as $\eta = (I - A)/2$, such that in principle the electronegativity and its variation with charge can be estimated for every atom. This also was the choice of many authors (see for example Refs. 13, 16, 45). For the atom in the molecule, this is best approached by the valence state ionization potentials[2, 53] and electron affinities, and was preferred by others (see for example Refs. 2, 20, 30, 54–56). This meets with the obvious difficulty that there is no general criterion for establishing the actual hybridization state of an atom or an ion in a molecule[17]. There also is a slight departure from a simple quadratic relation, which seems to be decreasing from the group I to the group VII elements[28]. Either one of the following functions seems to be appropriate[28, 57]:

$$E(N) = (aN + bN^2)\exp(cN^2) \tag{32a}$$

$$E(N) = (aN + bN^2)\exp(c|N|) \tag{32b}$$

where N is the number of electrons relative to the neutral atom for which $N = 0$. The parameters for the first and second row elements, obtained from a fit of the valence state ionization potentials (see Bash, Viste and Gray[58]) are given in Ref. 28. The difficulty of applying these concepts however, is the lack of accurate values for the valence state electron affinities (for a compilation, see e.g. Hotop and Lineberger[59]). Notice also that dE/dN and d^2E/dN^2 for Eq. (32b) are different for $N < 0$ or $N > 0$. There are of course also the calculated electronegativity scales (see at other places in this book) and the empirical scales such as the Sanderson scale[9, 10, 14] and the original Pauling scale. The latter scale was applied recently for the calculation of partial charges by Bratsch[47, 48].

For an extended discussion of the use of the orbital electronegativity scale by Hinze et al.[20, 54–56], Sanderson's scale[9] and a scale based on Eq. (32a), with the PEOE, FEOE formalism and the use of the average compound electronegativity, see Ref. 28. The choice of the electronegativity scale only influences the numerical values of the partial charges, but does barely alter the correlations with experimental data. Moreover, the choice of an empirical scale (Sanderson's scale e.g.) has the advantage that no choice of the hybridization state is required. As already pointed out before, the calibration of $\Delta\chi$ and $\Delta\eta$ in the FEOE method [Eq. (24)] already accounts for differences in the scale. Even when using a simple equalization of the isolated-atom electronegativities, χ^0 and η^0 may be adjusted so as to develop an internally consistent scheme, as it was done by Jardine et al.[46] to account for a change in the percentage of p character and to estimate the phenyl-group electronegativity. Many excellent correlations (although not always linear) comparing the different electronegativity scales were also presented in the literature, such that it is not at all surprising that any electronegativity scale can be used to obtain chemically significant information.

5. Electronegativity Equalization and Chemical Information

The beauty and simplicity of the application of the electronegativity equalization principles lies in its chemical intuition and the way in which chemical data can be rationalized quantitatively. After all, molecules are built from atoms, which retain some of their properties[12)], and no other "number" identifies an atom better than its electronegativity.

The concept of substituent electronegativity has been successfully used in physical organic chemistry, and has been touched upon at several places in this review. The reader is referred to the original references (e.g. Refs. 42, 46, 50, 52, 60) and excellent reviews of the subject (e.g. Ref. 61).

In Sect. 2, a variety of methods were discussed by which binding energies were successfully calculated. All of these rely in some way or another on the electronegativity equalization for the calculation of charges. These could then be used to calculate directly the contribution of the "electronegativity" energy [Eq. (10)] and the electrostatic energy to the total energy, and/or be used as a blending coefficient for estimating the relative importance of the electrostatic energy and the covalent part. In none of these methods does the "electronegativity energy" as such account for the total energy of formation. The most successful approach for calculating binding energies was made by Sanderson[9, 10)], who does not consider any contribution of the electronegativity energy at all.

Parr and Pearson[13)] used the charge shifts calculated by Eq. (29) to estimate the energy change in bond formation due to the change in "electronegativity" energy, as a basis for a quantitative expression of the hard and soft acid and base (HSAB) principle:

$$\Delta E = -\frac{1}{4}\frac{(\chi_\alpha^0 - \chi_\beta^0)^2}{(\eta_\alpha + \eta_\beta)} \tag{33}$$

Energy lowering follows from electron transfer. If the denominator in Eq. (33) is small, and for a reasonable difference in electronegativities, ΔE is substantial and stabilizing. This in part explains the HSAB principle: soft prefers soft. This does not explain, however, that hard prefers hard since in this case there is little electron transfer for a given difference in electronegativity. As already emphasized, the electronegativity energy does not account for the total energy of formation. By inclusion of the other terms in the expansion of the energy [Eq. (20)] Nalewajski[25)] extended the discussion of the HSAB principle to hard-hard and hard-soft cases.

The previously discussed relation between the average effective electronegativity and the average isolated-atom electronegativities for homologous series of compounds explains the success of many correlations between experimental properties and the average compound electronegativity (or the partial charges, by whatever method these may have been obtained). The former examples might create a false impression that the methods used here are only applicable to organic chemistry. On the contrary: this was, for example, the only way to quantitatively estimate compositional effects on the physicochemical properties of such complex systems as zeolites[62–64)]. It is, however, not necessarily so that for an extended range of the average electronegativity, the correlations should be linear[65)]. It was also found[66, 67)] that for the interaction of molecules with the acid sites of zeolites and silica-alumina, that the O–H bond was increasingly sensitive to perturbations (in the form of changes in composition) with decreasing bond strength.

The rapid calculation of connectivity-dependent charges using such formalisms as the modified Sanderson method[39], and PEOE and FEOE formalisms[28] may find multiple applications for rationalizing physicochemical properties based on the intrinsic properties of the atoms in a molecule. This was already demonstrated e.g. for ESCA chemical shifts[28, 30, 39] and resonance shifts in NMR[68]. The correct and rapid calculation of charges using the PEOE formalisms became part of the decision strategy in the synthetic design program EROS[69]. The projected use of the FEOE method for studying molecular interactions was already discussed before.

The average compound electronegativity, or the partial connectivity dependent charges may however not always contain sufficient information to explain the chemical properties. This is certainly so for binding energies (see above). Other molecular properties are also of importance, such as the polarizability. Huheey[70], discussing the gas-phase acidities of alcohols, came to the conclusion that two parameters are necessary to consider the effect of substituent groups: i.e. the "inherent electronegativity" and the relative "charge capacity". The latter parallels the hardness concept for the atom and also represents a fundamental molecular quantity. Gasteiger and Hutchings[71, 72] were able to derive a quantitative two-parameter model for gas-phase proton transfer reactions involving alcohols, ethers, thio-ethers and amines. Their model features an empirical model for the substituent polarizability[73], based on graph theory, and a composite "residual electronegativity" (PEOE formalism) including the charge-dependent residual electronegativities of the atoms closest to the reaction center (two bond spheres). Excellent correlations were found. There is no doubt that the residual electronegativity parallels Huheey's "inherent electronegativity" or the average molecular electronegativity (a quantity not produced by the PEOE formalism), and that their polarizability model stands for the "charge capacity" or molecular hardness. Hardness is not "equalized", and depends on the region within the molecule, and reflects the charge delocalization probabilities from the reaction center. In a way, residual charges also reflect the delocalization of the charges and can be used as a measure for the molecular hardness. These residual charges, together with the average molecular electronegativity (neither one of these being a sufficient parameter) do indeed explain the variation in the gas-phase protonation energies of amines, alkoxide ions and ethers[74]. The average electronegativity and the partial charges contain therefore important information, both of which can be considered as fundamental properties of the molecules.

6. Prospects

Electronegativity relations and the principles of electronegativity equalization will continue to be valuable tools in the hand of the chemist. This is even more evident after its rigorous definition in density-functional theory[4], which explains the revival of interest in the last few years. The concepts of electronegativity, partial charges, average electronegativity, atomic and molecular hardness are fundamental quantities. They are closely related to long-standing chemical ideas and can be quantitatively evaluated by different formalisms, especially in those areas where quantum-mechanical methods are not yet feasible. New developments in density-functional theory, such as a reactivity index, i.e. the Fukui function or frontier function defined by Parr and Yang[75] may in the future also lead to new simple formalisms for estimating molecular properties of interest to chemists.

7. Note Added in Proof

FEOE has since been reserved for the empirical formalism[28] only, and replaced by EEM (Electronegativity Equalization Method[29]). The EEM method has been successfully applied to the solid state: Van Genechten, K., Mortier, W. J., Geerlings, P.: J. Chem. Soc., Chem. Commun. 1278 (1986) and J. Chem. Phys. (1987, in press).

Acknowledgements. The author thanks the Belgian National Fund for Scientific Research (Nationaal Fonds voor Wetenschappelijk Onderzoek) for a permanent research position as Senior Research Associate (Onderzoeksleider) and NATO for a fellowship enabling a stay at the University of North Carolina at Chapel Hill. The active interest, discussions and suggestions by Prof. R. G. Parr, Dr. S. K. Ghosh and Dr. S. Shankar are gratefully acknowledged.

8. References

1. Pauling, L.: The Nature of the Chemical Bond, 3rd ed., Ithaca, New York, Cornell University Press 1960
2. Iczkowski, R. P., Margrave, J. L.: J. Am. Chem. Soc. *83,* 3547 (1961) (The association of the electronegativity with the first derivative of the total electronic energy with respect to the charge q, and its equivalence with Mulliken's definition of electronegativity, was first mentioned by Pritchard, H. O., Sumner, F. H.: Proc. Roy. Soc. London *A 235,* 136 (1956), p. 139 and Fig. 1.)
3. Mulliken, R. S.: J. Chem. Phys. *2,* 782 (1934)
4. Parr, R. G., Donnelly, R. A., Levy, M., Palke, W. E.: ibid. *68,* 3801 (1978)
5. Donnelly, R. A., Parr, R. G.: ibid *69,* 4431 (1978)
6. Parr, R. G.: Electron Distributions and the Chemical Bond (Coppens, P., Hall, M. B., eds.) p. 95, New York, Plenum Publ. Corp. 1982
7. Parr, R. G.: Ann. Rev. Phys. Chem. *34,* 631 (1983)
8. Hohenberg, P., Kohn, W.: Phys. Rev. B*136,* 864 (1964)
9. Sanderson, R. T.: Chemical Bonds and Bond Energy, New York, Academic Press 1976
10. Sanderson, R. T.: Polar Covalence, New York, Academic Press 1983
11. Politzer, P. Weinstein, H.: J. Chem. Phys. *71,* 4218 (1979)
12. Parr, R. G.: Int. J. Quant. Chem. *26,* 687 (1984)
13. Parr, R. G., Pearson, R. G.: J. Am. Chem. Soc. *105,* 7512 (1983)
14. Sanderson, R. T.: ibid. *105,* 2259 (1983)
15. Matcha, R. L.: ibid. *105,* 4859 (1983)
16. Evans, R. S., Huheey, J. E.: J. Inorg. Nucl. Chem. *32,* 777 (1970)
17. Ferreira, R.: Trans. Far. Soc. *59,* 1064 (1963)
18. Ferreira, R.: ibid. *59,* 1075 (1963)
19. Ponec, R.: Theoret. Chim. Acta (Berl.) *59,* 629 (1980)
20. Hinze, J., Whitehead, M. A., Jaffé, H. H.: J. Am. Chem. Soc. *85,* 148 (1963)
21. Reed, J. L.: J. Phys. Chem. *85,* 148 (1981)
22. Nethercot, A. H.: Chem. Phys. *59,* 297 (1981)
23. Komorowski, L.: Chem. Phys. Lett. *103,* 201 (1983)
24. Balbás, L. C., Alonso, J. H., Las Heras, E.: Mol. Phys. *48,* 981 (1983)

25. Nalewajski, R. F.: J. Am. Chem. Soc. *106,* 944 (1984)
26. Nalewajski, R. F., Koninski, M.: J. Phys. Chem. *88,* 6234 (1984)
27. Nalewajski, R. F.: ibid. *89,* 2831 (1985)
28. Mortier, W. J., Van Genechten, K., Gasteiger, J.: J. Am. Chem. Soc. *107,* 829 (1985)
29. Mortier, W. J., Ghosh, S. K., Shankar, S.: ibid. *108,* 4315 (1986)
30. Gasteiger, J., Marsili, M.: Tetrahedron *36,* 3219 (1980)
31. Hehre, W. J., Pople, J. A.: J. Am. Chem. Soc. *92,* 2191 (1970)
32. Brundle, C. R., Robin, M. B., Bash, H.: J. Chem. Phys. *53,* 2196 (1970)
33. Gutmann, V.: The Donor-Acceptor Approach to Molecular Interactions, New York, Plenum Press 1978
34. Gutmann, V.: Pure Appl. Chem. *51,* 2197 (1979)
35. Kolmann, P. A., Allen, L. C.: J. Phys. Chem. *51,* 3286 (1969)
36. Marsili, M., Gasteiger, J.: J. Croat. Chem. Acta *53,* 601 (1980)
37. Jolly, W. L., Perry, W. B.: J. Am. Chem. Soc. *95,* 5442 (1973)
38. Craver, J. C., Gray, R. C., Hercules, D. M.: ibid *96,* 6851 (1974)
39. Gray, R. C., Hercules, D. M.: J. Electron Spectr. Relat. Phen. *12,* 37 (1977)
40. Linert, W., Gutmann, V., Perkins, P. G.: Inorg. Chim. Acta *69,* 61 (1983)
41. Mortier, W. J.: unpublished results
42. Huheey, J. E.: J. Phys. Chem. *70,* 2086 (1966)
43. Parr, R. G., Bartolotti, L.: J. Am. Chem. Soc. *104,* 3801 (1982)
44. Perdew, J. P., Parr, R. G., Levy, M., Balduz, J. L.: Phys. Rev. Lett. *49,* 1691 (1982)
45. Huheey, J. E.: J. Phys. Chem. *69,* 3284 (1965)
46. Jardine, W. K., Langler, R. F., MacGregor, J. A.: Can. J. Chem. *60,* 2069 (1982)
47. Bratsch, S. G.: J. Chem. Educ. *61,* 588 (1984)
48. Bratsch, S. G.: ibid. *62,* 101 (1985)
49. Komorowski, L.: Chem. Phys. *76,* 31 (1983)
50. Inamoto, N., Masuda, S.: Chem. Lett. 1003 (1982)
51. Gordy, W.: Phys. Rev. *69,* 604 (1946)
52. Mullay, J.: J. Am. Chem. Soc. *106,* 5842 (1984)
53. McGlinn, S. P., Van Quickenborne, L. C., Kinoshita, M., Caroll, D. G.: Introduction to Applied Quantum Chemistry, San Francisco, Holt, Rinehart and Winston, 1972, p. 106
54. Hinze, J., Jaffé, H. H.: J. Chem. Phys. *84,* 540 (1962)
55. Hinze, J., Jaffé, H. H.: Can. J. Chem. *41,* 1315 (1963)
56. Hinze, J., Jaffé, H. H.: J. Phys. Chem. *67,* 1501 (1963)
57. Mortier, W. J.: unpublished results
58. Bash, H., Viste, A., Gray, H. B.: Theor. Chim. Acta *3,* 458 (1965)
59. Hotop, H., Lineberger, W. C.: J. Phys. Chem. Ref. Data *4,* 539 (1975)
60. Reynolds, W. F., Taft, R. W., Marriott, S., Topsom, R. D.: Tetrahedron Lett. *23,* 1055 (1982)
61. Wells, P. R.: Progr. Phys. Org. Chem. *6,* 111 (1968)
62. Mortier, W. J.: J. Catal. *55,* 13 (1978)
63. Jacobs, P. A., Mortier, W. J., Uytterhoeven, J. B.: J. Inorg. Nucl. Chem. *40,* 1919 (1978)
64. Jacobs, P. A., Mortier, W. J.: Zeolites *2,* 226 (1982)
65. Lercher, J. A., Noller, H.: J. Catal. *77,* 152 (1982)
66. Geerlings, R., Tariel, N., Botrel, A., Lissillour, R., Mortier, W. J.: J. Phys. Chem. *88,* 5752 (1984)
67. Datka, J., Geerlings, P., Mortier, W. J., Jacobs, P. A.: ibid. *89,* 3448 (1985)
68. Gasteiger, J., Marsili, M.: Org. Magn. Res. *15,* 353 (1980)
69. Gasteiger, J., Jochum, C.: Top. Curr. Chem. *74,* 93 (1978)
70. Huheey, J. E.: J. Org. Chem. *36,* 204 (1971)
71. Gasteiger, J., Hutchings, M. G.: J. Am. Chem. Soc. *106,* 6489 (1984)
72. Hutchings, M. G., Gasteiger, J.: Tetrahedron Lett. *24,* 2541 (1983)
73. Gasteiger, J., Hutchings, M. G.: ibid. *24,* 2537 (1983)
74. Van Genechten, K., Uytterhoeven, L., Mortier, W. J.: in preparation
75. Parr, R. G., Yang, W.: J. Am. Chem. Soc. *106,* 4049 (1984)

Electronegativity and Charge Distribution

D. Bergmann and Juergen Hinze

Fakultät für Chemie, Universität Bielefeld, D-4800 Bielefeld, FRG

The concept of orbital electronegativity is extended such that the orbital electronegativity does depend as well on the hybridization as on the partial charge of the atom. The parameters to determine these orbital electronegativity values can be and have been obtained from spectroscopic data of the atoms and corresponding ions. With this and the concept of electronegativity equalization in a localized 2-electron bond, it is possible to determine the partial charges on the atoms in a molecule, by merely solving a set of linear equations with the dimension equal to the number of bonds in the molecule. Similarly it is possible to compute the orbital electronegativities of molecular fragments. Such group electronegativities and partial charges computed for a large number of molecules correlate well with a number of experimentally observable quantities.

The concept of electronegativity, as identified by Parr with the chemical potential being uniform throughout an entire molecule, is analysed quantum mechanically.

A. Introduction 146

B. Orbital-Electronegativity and Hardness 147

C. Electronegativity Equalization 149

D. Multi-Bonded Atoms: The Charge Distribution in Molecules 150

E. Group Electronegativities and Hardness 164

F. Correlations with Molecular Properties 168

I. Bond Length 168

II. NMR-Chemical Shifts 171

III. Proton Affinities 172

Appendix A 181

Appendix B 188

G. References 189

Structure and Bonding 66

A. Introduction

The intuitively appealing concept of electronegativity as "the power of an atom in a molecule to attract charge to itself" gained rapid popularity among chemists after its first definite introduction by Pauling[1]. This popularity has hardly decreased, even though, or because, electronegativity could not be derived rigorously from the principles of quantum mechanics. In conjunction with electronegativity, equally flexible concepts, like electron donating or withdrawing groups, partial charges and electrophilic and nucleophilic centres in molecules are used effectively in rationalizing and predicting qualitatively the characteristics and reactivities of molecules.

These concepts not rigidly fixed by a derivation from first principles are somewhat ambiguous, though th3y are powerful, for they may be adjusted within narrow limits in the light of new experimental information and theoretical insight, a necessity until we understand chemi try completely.

Many molecular properties like bond distances, force constants, ESCA K-shell energies, proton affinities, NMR chemical shifts and coupling constants etc. correlate well with electronegativities of the atoms or data easily derived with these electronegativities. Thus, in turn the atomic electronegativities can be used to predict or estimate such molecular properties with ease[2]. To be sure, all these molecular properties could be obtained via direct quantum mechanical calculations, a task often much too arduous.

The large number of molecular properties which correlate with the atomic electronegativities have led to a number of electronegativity scales defined in terms of such molecular properties[3]; this, however, is fallacious. It does not contribute to the desired aim to understand and derive molecular characteristics in terms of the measureable properties of the constituent atoms. Equally faulty is the following argument: if molecular properties A, say force constants, correlate with electronegativity and properties B, say bond distances, correlate with electronegativity also, then obviously the molecular properties A and B will correlate, thus there is no need for the (ill defined) intermediate atomic electronegativity[4]. The point missed, is the data reduction desired and achieved, if we reach the goal of understanding and predicting molecules and their properties on the basis of their constituent atoms. There are several million different molecules and only 100 different atoms!

Recently, a seemingly rigorous derivation of atomic and molecular electronegativity has been presented on the basis of the density-functional theory[5], with electronegativity defined as the partial derivative of the total energy with respect to the total number of electrons, the trace of the charge density. This definition, restricted to the ground state of a system, as

$$\chi = -\left(\frac{\partial E}{\partial N}\right)_{v_{(1)}} \tag{1}$$

with $v_{(1)}$ the one particle potential fixed is akin to the earlier definitions of the electronegativity by Iczkowski and Margrave[6] and of orbital-electronegativity by Hinze et al.[7]. We will elaborate here on this orbital electronegativity concept, which can be derived rigorously and appears to be more in the spirit of the traditional meaning of electronegativity, so useful in chemistry. In Appendix A we will discuss the connections and differences between this more traditional concept and the results obtained using density functional theory.

The density functional derivation of electronegativity has sparked off renewed theoretical interest in electronegativity-related concepts[8], as is testified to by this volume. This is in particular so, because the density functional derivation seems also to justify the assumption of electronegativity equalization in a bond, as originally proposed by Sanderson[9], used effectively and elaborated on by many authors[10].

In the following, we will present a clear definition of the orbital electronegativity for the valence states of atoms. These can be determined directly from the known atomic spectroscopic data. With this, in conjunction with the concept of electronegativity equalization in a bond, ionicity and partial charges in a molecule can be obtained. As the dependence of orbital electronegativity on such partial charges can be determined also from atomic spectroscopic data, we will describe how charge distributions in molecules can be calculated easily. With this, group electronegativities in general and hardness parameters[11] for acids and bases, as well as electro- and nucleophiles become readily accessible, all derived directly from atomic spectroscopic data. We will conclude by correlating the results obtained with various measured molecular properties to emphasize the usefulness of the concepts presented as a means to predict such data.

B. Orbital-Electronegativity and Hardness

More then fifty years ago, Mulliken[12] had defined the electronegativity of an atom as

$$\chi = \frac{I + EA}{2} \tag{2}$$

and realized that the ionization potential I and the electron affinity EA used should not be those of an atom in its ground state. As electronegativity is to be the power (potential) of an atom in a molecule to attract electrons to itself, the ionization potential and electron affinity of the valence-states atoms assume in a molecule are more appropriate. To be sure, the specification and characterization of the particular valence state an atom assumes in a molecule is problematic. Good approximations are often the sp^3 (te = tetrahedral) hybrids or the sp^2 (tr = trigonal) or sp (di = diagonal) hybrids; greater difficulties arise if d-orbital involvement is significant with the extension of the octet rule. For defined valence states it is possible to determine the ionization potentials and electron affinities needed in Eq. (2) directly from the experimentally known atomic term values[13].

We focus attention now on a particular atomic hybrid orbital (a spatial orbital), which will be used for bond formation, and define its electronegativity as[14]

$$\chi_i(q_i) = \left(\frac{\partial E}{\partial q_i}\right)_{n^0} = -\left(\frac{\partial E}{\partial n_i}\right) \tag{3}$$

with E the energy of the atom and q_i the negative of the occupation number, n_i of, i.e. the charge in, orbital i considered. This charge is treated as a continuous variable, $0 > q_i >$

-2. The index n^0 is to signify that all the other orbital occupations are to be held fixed. A rigorous quantum mechanical justification of this definition, Eq. (3), will be given in Appendix A.

With the energy of the atom approximated as

$$E(q_i) = a_i + b_i q_i + c_i q_i^2 \tag{4}$$

we obtain for the orbital electronegativity

$$\chi_i(q_i) = b_i + 2\,c_i q_i \tag{5}$$

As the electronegativity is to represent the potential to attract electrons before bond formation of the orbitals, the values for q_i equal to 0, -1 and -2 attain special importance.

With the valence-state ionization potential and electron affinity of the orbital i defined as

$$\begin{aligned} I_i &= E_i(0) - E_i(-1) \\ EA_i &= E_i(-1) - E_i(-2) \end{aligned} \tag{6}$$

we obtain with Eq. (4), where we have dropped the index i,

$$\begin{aligned} b &= (3\,I - EA)/2 \\ c &= (I - EA)/2 \end{aligned} \tag{7}$$

This yields for the electronegativity of a singly occupied orbital, corresponding to the standard electronegativity

$$\chi(-1) = (I + EA)/2 \tag{8}$$

in accord with the original definition of Mulliken.

For empty and doubly occupied orbitals we obtain respectively

$$\begin{aligned} \chi(0) &= (3\,I - EA)/2 \\ \chi(-2) &= (3\,EA - I)/2 \end{aligned} \tag{9}$$

The orbital electronegativities obtained in this way have the dimension of an electrical potential, the same as the parameter b. With the ionization energy and electron affinity given in eV the units for χ and b will be volts. To convert the results to the "Pauling units" more familiar to chemists, the empirical relation[15)]

$$\chi_p = 0.336\,(\chi_M - 0.615) \tag{10}$$

may be used, where χ_p is the electronegativity in "Pauling units" while χ_M (Mulliken) is in volts.

The hardness parameter of an orbital defined as

$$\eta_i = \frac{1}{2}\left(\frac{\partial^2 E}{\partial q_i^2}\right)_{n_0} = \frac{1}{2}\frac{\partial \chi_i}{\partial q_i} \tag{11}$$

is obtained directly from Eq. (5), i.e. we have

$$\eta_i = c_i \tag{12}$$

It has the dimensions of energy/charge^2 or potential/charge which is the same as capacitance^{-1}. With the units for energy as eV used here its units will be V/e.

C. Electronegativity Equalization

With the orbital electronegativity defined in Eq. (3) a rationalization of the electronegativity equalization between the two orbitals forming a two-electron bond as suggested by Sanderson[9] appears to be immediate. If an orbital i on atom A say forms a bond with orbital j on atom B, a transfer of charge Δq will occur from the less electronegative orbital to the more electronegative one, until the electronegativities

$$\chi_i(q_i^0 + \Delta q) = \chi_j(q_j^0 - \Delta q) \tag{13}$$

have become equal. This charge transfer process, until the electronegativities are equalized, would correspond to an energy lowering in the bond, provided, everything else would remain equal, i.e. there would be no other energetic effects connected with this charge transfer.

To be sure, there will be other energetic effects in case of such a charge transfer; these have been discussed extensively[10]. It is fortunate that the two main effects: (i) an additional energy lowering, due to the Coulomb attraction between the now partially positive and negative centres A and B, and (ii) an energy increase because of a weakening of the covalent bond, have an opposite direction and therefore will tend to cancel. For dative bonds between neutral species, i.e. if $q_i^0 = 0$ and $q_j^0 = -2$ the effect (ii) would be energy lowering also resulting in a larger charge transfer. Thus the assumption of a full electronegativity equalization in bond formation appears to be not bad and we will adhere to it here.

The charge transferred from orbital j to orbital i is obtained immediately by using Eq. (13) with Eqs. (5) and (7) as

$$\Delta q = \frac{\chi_j(q_j^0) - \chi_i(q_i^0)}{2(c_j + c_i)} = \frac{\chi_j(q_j^0) - \chi_i(q_i^0)}{I_j - EA_j + I_i - EA_i} \tag{14}$$

and the net energy decrease due to this charge transfer will be

$$\Delta E = \int_0^{\Delta q} \left(\frac{\partial E_i}{\partial q_i}\right)_{q_i^0+q} - \left(\frac{\partial E_j}{\partial q_j}\right)_{q_j^0-q} dq \tag{15}$$

$$= \int_0^{\Delta q} [\chi_i(q_i^0 + q) - \chi_j(q_j^0 - q)]dq$$

This gives with Eqs. (5) and (14) the extra-ionic resonance energy

$$\Delta E = -\frac{[\chi_j(q_j^0) - \chi_i(q_i^0)]^2}{4(c_j + c_i)} \tag{16}$$

Pauling[1)] used to establish the first scale of relative electronegativity values; the square of their difference was to be proportional to the extra-ionic resonance energy.

The discussion above applies to a single localized two-electron bond only. Chemistry would be pretty dull if atoms could engage in one bond only.

D. Multi-Bonded Atoms: The Charge Distribution in Molecules

If we are to extend the concepts of electronegativity and its equalization in bonds to atoms engaging in more than one bond in a molecule, an extension of these concepts is necessary. Consider an atom, using m orbitals to form m localized two-electron bonds, then due to the orbital-electronegativity equalization in each bond we will have a charge transfer of Δq_k into each of the atomic orbitals, i.e. for $k = 1$, through m. The total net charge of the originally neutral atom will be

$$Q = \sum_{k=1}^{m} \Delta q_k \tag{17}$$

If we focus now on orbital i and its electronegativity χ_i we need this not only as a function of $q_i = q_i^0 + \Delta q_i$ as given by Eq. (5), but we will have to consider also the rest charge of the atom due to the charge transfer in all the other bonds of the atom in the molecule. This rest charge is

$$r_i = \sum_{k \neq i} \Delta q_k = Q - \Delta q_i \tag{18}$$

Certainly the orbital electronegativity χ_i will depend on this rest-charge and we have to generalize Eq. (5) to

$$\chi_i(r_i, q_i) = b_i(r_i) + 2\,c_i(r_i)q_i \tag{19}$$

with the parameters b and c dependent on the rest charge. Fortunately the parameter $c = (I - EA)/2$ depends only weakly on r and we will treat it as being constant as r will in general be small; i.e. $|r| < 1$. For the r-dependence of the parameter b we will use a linear approximation

$$b(r) = b^0 + b^1 r \tag{20}$$

with

$$b = (3I - EA)/2 = I + c \tag{21}$$

we have

$$b^0 = I^0 + c \tag{22}$$

and

$$b^1 = \frac{\partial I}{\partial r} \tag{23}$$

The parameter b^1 can be evaluated for the various valence orbitals analogously to the b^0 and c values, using valence-state promotion energies together with the ground state ionization potentials and electron affinities[7)]. There is one minor complication, however Slightly different b^1 values will be obtained, depending on the orbitals out of which the rest charge originates. As this dependence would unduely complicate the procedure proposed below to calculate the charge distribution, and since this dependence is only small, we have determined suitably averaged b^1 values. The b's and c's obtained[16)] for various valence orbitals, using spectroscopic data[17)], of the atoms are listed in Table 1, together with the corresponding orbital electronegativities.

We are now in a position to give explicitly the orbital electronegativities as they depend on the charge, q_i, in orbital i and on the rest-charge, r_i, defined in Eq. (18). The result is Eq. (24),

$$\chi_i(r_i, q_i) = b_i^0 + b_i^1 r_i + 2\,c_i q_i \tag{24}$$

which may now be used to calculate the charge shift Δq_i obtained because of electronegativity equalization if orbital i forms a localized two-electron bond with orbital j.

$$\chi_i(r_i, q_i^0 + \Delta q_i) = \chi_j(r_j, q_j^0 + \Delta q_j) \tag{25}$$

with

$$\Delta q_j = -\Delta q_i\,.$$

Note that the r_i and r_j's in Eq. (25) depend on the Δq's, see Eq. (18), in the other bonds of the two atoms involved. In a molecule with N bonds we will have N charge shifts, Δq, and N equations of type (25). Thus we obtain N linear equations for the determination of the N unknowns Δq_i, see Appendix B for the details. These equations are readily set up and solved for almost arbitrarily large molecules. Solving these equations leads directly with Eq. (17) to atomic charge distributions in the molecule. These charge distributions correlate well with those values obtained via a Mulliken population analysis of ab-initio Hartree-Fock results, see Table 2.

Table 1. Electronegativities and electronegativity parameters for the elements[a]

Atom	Valence state	b^0	b^1	c	χ_M	χ_P
H	s	20.021		6.422	7.18	2.20
Li	s	7.778		2.386	3.01	0.80
	p	5.139		1.594	1.95	0.45
Be	$\underline{s}\ p$	13.292	6.754	3.373	6.55	1.99
	$s\ \underline{p}$	8.891	7.222	2.930	3.03	0.81
	p p	8.791	6.350	2.678	3.43	0.95
	di di	12.376	6.345	3.794	4.79	1.40
	$\underline{di}\ p_\pi$	11.566	6.551	3.550	4.47	1.29
	$di\ \underline{p_\pi}$	8.841	6.668	2.804	3.23	0.88
	tr tr	11.130	6.437	3.513	4.10	1.17
	$\underline{tr}\ p_\pi$	10.759	6.485	3.377	4.01	1.14
	$tr\ \underline{p_\pi}$	8.826	6.536	2.763	3.30	0.90
	te te	10.525	6.450	3.338	3.85	1.09
B	$s^2\underline{p}$	12.721		4.423	3.88	1.10
	$\underline{s}\ p^2$	20.514		4.798	10.92	3.46
	$p^2\underline{p}$	11.875		2.666	6.54	1.99
	$di^2\underline{di}$	19.835		5.683	8.47	2.64
	$di^2\underline{p_\pi}$	12.992		3.861	5.27	1.56
	$\underline{di}\ p_\pi^2$	17.207		4.744	7.72	2.39
	$tr^2\underline{tr}$	17.226		5.030	7.17	2.20
	$tr^2\underline{p_\pi}$	12.774		3.533	5.71	1.71
	$\underline{tr}\ p_\pi^2$	15.654		4.276	7.10	2.18
	$te^2\underline{te}$	15.902		4.570	6.76	2.07
	$\underline{s}\ p\ p$	19.547	8.795	4.633	10.28	3.25
	$s\ \underline{p}\ p$	12.502	9.247	4.079	4.34	1.25
	p p p	10.905	7.233	2.499	5.91	1.78
	$\underline{di}\ di\ p_\pi$	17.791	8.594	5.239	7.31	2.25
	$di\ di\ \underline{p_\pi}$	12.154	9.247	3.920	4.31	1.24
	$\underline{di}\ p_\pi p_\pi$	16.237	8.013	4.577	7.08	2.17
	$di\ \underline{p_\pi}p_\pi$	11.704	8.303	3.289	5.13	1.52
	tr tr tr	16.266	8.717	4.978	6.31	1.91
	$\underline{tr}\ tr\ p_\pi$	15.552	8.264	4.579	6.39	1.94
	$tr\ tr\ \underline{p_\pi}$	11.815	8.633	3.482	4.85	1.42
	te te te	14.913	8.449	4.477	5.96	1.80
C	$s^2\underline{p}\ p$	16.562	11.663	5.619	5.32	1.58
	$\underline{s}\ p^2p$	27.548	12.070	6.112	15.32	4.94
	$s\ p^2\underline{p}$	17.225	11.669	5.529	6.17	1.87
	$p^2\underline{p}\ p$	19.974	11.345	6.540	6.89	2.11
	$di^2\underline{di}\ p_\pi$	26.160	12.024	7.234	11.69	3.72
	$di^2di\ \underline{p_\pi}$	16.894	11.743	5.574	5.75	1.72
	$\underline{di}\ di\ p_\pi^2$	24.962	10.583	7.108	10.75	3.40
	$di^2\underline{p_\pi}p_\pi$	18.271	11.340	5.921	6.43	1.95
	$\underline{di}\ p_\pi^2p_\pi$	24.807	11.707	7.372	10.06	3.17
	$di\ p_\pi^2\underline{p_\pi}$	18.600	11.267	6.035	6.53	1.99
	$tr^2\underline{tr}\ tr$	23.414	11.834	6.934	9.55	3.00
	$tr^2\underline{tr}\ p_\pi$	23.652	12.191	7.093	9.47	2.97
	$tr^2tr\ \underline{p_\pi}$	17.923	11.914	5.826	6.27	1.90
	$\underline{tr}\ tr\ p_\pi^2$	23.193	11.016	7.098	9.00	2.82

Table 1 (continued)

Atom	Valence state	b^0	b^1	c	χ_M	χ_P
C	$\mathrm{te^2\underline{te}\ te}$	22.260	11.323	6.836	8.59	2.68
	$\mathrm{\underline{s}\ p\ p\ p}$	26.997	11.591	5.977	15.04	4.85
	$\mathrm{s\ \underline{p}\ p\ p}$	16.675	11.092	5.394	5.89	1.77
	$\mathrm{\underline{di}\ di\ p_\pi p_\pi}$	24.409	11.020	6.972	10.47	3.31
	$\mathrm{di\ di\ \underline{p}_\pi p_\pi}$	16.672	11.119	5.473	5.73	1.72
	$\mathrm{\underline{tr}\ tr\ tr\ p_\pi}$	22.400	10.984	6.766	8.87	2.77
	$\mathrm{tr\ tr\ tr\ \underline{p}_\pi}$	16.671	11.128	5.499	5.67	1.70
	$\mathrm{te\ te\ te\ te}$	21.185	10.996	6.565	8.06	2.50
N	$\mathrm{s^2p^2\underline{p}}$	20.499		6.437	7.62	2.36
	$\mathrm{\underline{s}\ p^2p^2}$	33.368		6.323	20.72	6.76
	$\mathrm{p^2p^2\underline{p}}$	18.843		4.658	9.53	2.99
	$\mathrm{di^2di^2\underline{p}_\pi}$	20.499		6.437	7.62	2.36
	$\mathrm{di^2\underline{di}\ p_\pi^2}$	32.147		8.118	15.91	5.14
	$\mathrm{di^2p_\pi^2\underline{p}_\pi}$	20.209		5.461	9.29	2.91
	$\mathrm{\underline{di}\ p_\pi^2p_\pi^2}$	27.132		6.517	14.10	4.53
	$\mathrm{tr^2tr^2\underline{tr}}$	29.421		7.943	13.53	4.34
	$\mathrm{tr^2tr^2\underline{p}_\pi}$	20.425		5.767	8.89	2.78
	$\mathrm{tr^2\underline{tr}\ p_\pi^2}$	27.247		6.995	13.26	4.25
	$\mathrm{te^2te^2\underline{te}}$	26.434		7.046	12.34	3.94
	$\mathrm{s^2\underline{p}\ p\ p}$	20.601	13.263	6.647	7.31	2.25
	$\mathrm{\underline{s}\ p^2p\ p}$	33.473	12.638	6.534	20.41	6.65
	$\mathrm{s\ p^2\underline{p}\ p}$	20.436	13.095	6.004	8.43	2.62
	$\mathrm{di^2\underline{di}\ p_\pi p_\pi}$	32.248	13.249	8.327	15.59	5.03
	$\mathrm{di^2di\ \underline{p}_\pi p_\pi}$	20.518	13.172	6.325	7.87	2.44
	$\mathrm{\underline{di}\ di\ p_\pi^2p_\pi}$	29.804	12.331	7.694	14.42	4.64
	$\mathrm{di\ di\ p_\pi^2\underline{p}_\pi}$	20.168	13.095	6.047	8.07	2.51
	$\mathrm{tr^2\underline{tr}\ tr\ p_\pi}$	28.419	12.890	7.800	12.82	4.10
	$\mathrm{tr^2tr\ tr\ \underline{p}_\pi}$	20.372	13.144	6.238	7.90	2.45
	$\mathrm{\underline{tr}\ tr\ tr\ p_\pi^2}$	27.195	12.467	7.466	12.26	3.91
	$\mathrm{te^2\underline{te}\ te\ te}$	26.399	12.826	7.456	11.49	3.65
O	$\mathrm{s^2p^2\underline{p}\ p}$	24.929	15.137	7.647	9.63	3.03
	$\mathrm{\underline{s}\ p^2p^2p}$	44.900	15.998	8.826	27.25	8.95
	$\mathrm{s\ p^2p^2\underline{p}}$	26.110	14.997	7.572	10.97	3.48
	$\mathrm{di^2di^2\underline{p}_\pi p_\pi}$	24.929	15.137	7.647	9.63	3.03
	$\mathrm{di^2\underline{di}\ p_\pi^2p_\pi}$	40.159	16.166	9.985	20.19	6.58
	$\mathrm{di^2di\ p_\pi^2\underline{p}_\pi}$	25.520	15.327	7.610	10.30	3.25
	$\mathrm{\underline{di}\ di\ p_\pi^2p_\pi^2}$	38.327	14.086	9.610	19.11	6.21
	$\mathrm{tr^2tr^2\underline{tr}\ p_\pi}$	36.249	15.956	9.595	17.06	5.53
	$\mathrm{tr^2tr^2tr\ \underline{p}_\pi}$	25.323	15.323	7.622	10.08	3.18
	$\mathrm{tr^2\underline{tr}\ tr\ p_\pi^2}$	35.565	15.132	9.419	16.73	5.41
	$\mathrm{te^2te^2\underline{te}\ te}$	33.545	15.393	9.150	15.25	4.92
F	$\mathrm{s^2p^2p^2\underline{p}}$	29.595		8.726	12.14	3.87
	$\mathrm{\underline{s}\ p^2p^2p^2}$	45.234		6.981	31.27	10.30
Na	s	7.436		2.297	2.84	0.75
	p	4.470		1.434	1.60	0.33
Mg	$\mathrm{\underline{s}\ p}$	12.031	5.347	3.077	5.88	1.77
	$\mathrm{s\ \underline{p}}$	6.758	5.463	2.233	2.29	0.56
	$\mathrm{p\ p}$	8.476	4.808	2.823	2.83	0.74

Table 1 (continued)

Atom	Valence state	b^0	b^1	c	χ_M	χ_P
Mg	di di	10.119	5.035	3.015	4.09	1.17
	$\underline{di}\ p_\pi$	10.565	5.077	3.262	4.04	1.15
	$di\ \underline{p}_\pi$	7.617	5.084	2.528	2.56	0.65
	tr tr	9.548	5.019	3.009	3.53	0.98
	$\underline{tr}\ p_\pi$	9.939	4.988	3.186	3.57	0.99
	$tr\ \underline{p}_\pi$	7.903	4.981	2.626	2.65	0.68
	te te	9.272	4.989	2.985	3.30	0.90
Al	$s^2\underline{p}$	9.050		3.064	2.92	0.78
	$\underline{s}\ p^2$	16.983		3.931	9.12	2.86
	$p^2\underline{p}$	8.338		1.058	6.22	1.88
	$di^2\underline{di}$	14.960		4.144	6.67	2.04
	$di^2\underline{p}_\pi$	9.114		2.256	4.60	1.34
	$\underline{di}\ p_\pi^2$	13.282		3.116	7.05	2.16
	$tr^2\underline{tr}$	12.786		3.336	6.11	1.85
	$tr^2\underline{p}_\pi$	8.950		1.901	5.15	1.52
	$\underline{tr}\ p_\pi^2$	11.772		2.568	6.64	2.02
	$te^2\underline{te}$	11.686		2.849	5.99	1.81
	$\underline{s}\ p\ p$	15.985	6.302	3.709	8.57	2.67
	$s\ \underline{p}\ p$	9.055	6.514	2.583	3.89	1.10
	p p p	7.340	4.524	0.836	5.67	1.70
	$\underline{di}\ di\ p_\pi$	13.592	6.147	3.682	6.23	1.89
	$di\ di\ \underline{p}_\pi$	8.840	6.512	2.483	3.87	1.10
	$\underline{di}\ p_\pi p_\pi$	12.284	5.414	2.894	6.50	1.98
	$di\ \underline{p}_\pi p_\pi$	8.198	5.561	1.710	4.78	1.40
	tr tr tr	12.222	6.210	3.390	5.44	1.62
	$\underline{tr}\ tr\ p_\pi$	11.546	5.682	2.890	5.77	1.73
	$tr\ tr\ \underline{p}_\pi$	8.388	5.887	1.956	4.48	1.30
	te te te	10.996	5.854	2.826	5.34	1.59
Si	$s^2\underline{p}\ p$	11.109	7.488	3.439	4.23	1.21
	$\underline{s}\ p^2p$	22.711	8.303	5.200	12.31	3.93
	$s\ p^2\underline{p}$	12.596	7.996	3.204	6.19	1.87
	$p^2\underline{p}\ p$	14.859	4.948	2.657	9.54	3.00
	$di^2\underline{di}\ p_\pi$	19.386	7.968	5.145	9.10	2.85
	$di^2di\ \underline{p}_\pi$	11.870	7.803	3.327	5.22	1.55
	$\underline{di}\ di\ p_\pi^2$	19.269	7.341	5.010	9.25	2.90
	$di^2\underline{p}_\pi p_\pi$	12.422	5.664	2.451	7.52	2.32
	$\underline{di}\ p_\pi^2p_\pi$	18.978	6.625	4.122	10.73	3.40
	$di\ p_\pi^2\underline{p}_\pi$	13.727	5.857	2.930	7.87	2.44
	$tr^2\underline{tr}\ tr$	17.295	7.630	4.675	7.94	2.46
	$tr^2\underline{tr}\ p_\pi$	17.160	6.700	3.952	9.26	2.90
	$tr^2tr\ \underline{p}_\pi$	12.609	6.319	2.836	6.94	2.12
	$\underline{tr}\ tr\ p_\pi^2$	17.526	6.449	4.132	9.26	2.91
	$te^2\underline{te}\ te$	16.249	6.724	3.932	8.38	2.61
	$\underline{s}\ p\ p\ p$	22.531	8.026	5.223	12.09	3.85
	$s\ \underline{p}\ p\ p$	12.416	6.990	3.225	5.97	1.80
	$\underline{di}\ di\ p_\pi p_\pi$	19.090	7.443	5.032	9.03	2.83
	$di\ di\ \underline{p}_\pi p_\pi$	12.706	7.076	3.527	5.65	1.69
	$\underline{tr}\ tr\ tr\ p_\pi$	17.354	7.316	4.744	7.87	2.44

Table 1 (continued)

Atom	Valence state	b^0	b^1	c	χ_M	χ_P
Si	$tr\ tr\ tr\ \underline{p_\pi}$	12.804	7.108	3.628	5.55	1.66
	$te\ te\ te\ te$	16.373	7.266	4.557	7.26	2.23
P	$s^2p^2\underline{p}$	15.578		4.792	5.99	1.81
	$\underline{s}\ p^2p^2$	26.250		5.996	14.26	4.58
	$s^2p^2\ 2d$	4.899		1.826	1.25	0.21
	$s^2\underline{p}\ d^2$	26.313		6.685	12.94	4.14
	$\underline{s}\ p^2d^2$	32.991		6.398	20.20	6.58
	$s^2d^2\underline{d}$	16.076		3.660	8.76	2.74
	$\underline{s}\ d^2d^2$	40.462		7.004	26.45	8.68
	$p^2p^2\underline{p}$	24.191		7.445	9.30	2.92
	$p^2p^2\underline{d}$	12.977		4.421	4.13	1.18
	$p^2\underline{p}\ d^2$	30.394		7.789	14.82	4.77
	$p^2d^2\underline{d}$	19.520		4.607	10.31	3.26
	$\underline{p}\ d^2d^2$	37.228		8.239	20.75	6.77
	$d^2d^2\underline{d}$	26.764		4.965	16.83	5.45
	$di^2di^2\underline{p_\pi}$	15.578		4.792	5.99	1.81
	$di^2\underline{di}\ p_\pi^2$	24.019		6.429	11.16	3.54
	$di^2p_\pi^2\underline{p_\pi}$	21.708		6.754	8.20	2.55
	$\underline{di}\ p_\pi^2p_\pi^2$	25.703		7.203	11.30	3.59
	$tr^2tr^2\underline{tr}$	21.897		6.114	9.67	3.04
	$tr^2tr^2\underline{p_\pi}$	20.071		6.242	7.59	2.34
	$tr^2\underline{tr}\ p_\pi^2$	24.412		7.035	10.34	3.27
	$te^2te^2\underline{te}$	22.985		6.712	9.56	3.01
	$te^2te^2\underline{d_\pi}$	10.744		3.747	3.25	0.89
	$te^2\underline{te}\ d_\pi^2$	31.955		7.977	16.00	5.17
	$pl^2pl^2\underline{pl}$	23.723		6.213	11.30	3.59
	$pl^2pl^2\underline{p_\pi}$	25.109		6.938	11.23	3.57
	$pl^2\underline{pl}\ p_\pi^2$	23.062		6.452	10.16	3.21
	$tp^2tp^2\underline{tp}$	24.331		5.550	13.23	4.24
	$pe^2pe^2\underline{pe}$	23.810		6.483	10.84	3.44
	$qp^2qp^2\underline{qp}$	23.308		5.599	12.11	3.86
	$oh^2oh^2\underline{oh}$	24.649		6.396	11.86	3.78
	$s^2\underline{p}\ p\ p$	15.419	8.413	4.686	6.05	1.83
	$\underline{s}\ p^2p\ p$	26.090	12.528	5.889	14.31	4.60
	$s\ p^2\underline{p}\ p$	17.759	9.872	5.270	7.22	2.22
	$s^2\underline{p}\ p\ d$	20.767	7.127	5.714	9.34	2.93
	$s^2p\ p\ \underline{d}$	4.862	5.141	1.842	1.18	0.19
	$\underline{s}\ p^2p\ d$	29.444	10.728	6.173	17.10	5.54
	$s\ p^2\underline{p}\ d$	21.862	8.448	5.798	10.27	3.24
	$s\ p^2p\ \underline{d}$	6.797	6.959	2.259	2.28	0.56
	$\underline{s}\ p\ p\ d^2$	32.954	10.902	6.414	20.13	6.56
	$s\ \underline{p}\ p\ d^2$	26.212	9.119	6.373	13.47	4.32
	$s^2\underline{p}\ d\ d$	26.200	5.686	6.696	12.81	4.10
	$s^2p\ \underline{d}\ d$	10.325	3.553	2.734	4.86	1.43
	$\underline{s}\ p^2d\ d$	32.879	8.884	6.408	20.06	6.53
	$s\ p^2\underline{d}\ d$	11.102	5.295	2.740	5.62	1.68
	$\underline{s}\ p\ d^2d$	36.564	9.047	6.692	23.18	7.58
	$s\ \underline{p}\ d^2d$	30.662	7.590	6.985	16.69	5.40

Table 1 (continued)

Atom	Valence state	b^0	b^1	c	χ_M	χ_P
P	$s\ p\ d^2\underline{d}$	15.657	5.980	3.269	9.12	2.86
	$s^2\underline{d}\ d\ d$	15.963	1.919	3.670	8.62	2.69
	$\underline{s}\ d^2d\ d$	40.349	7.147	7.014	26.32	8.64
	$s\ d^2\underline{d}\ d$	20.312	4.359	3.836	12.64	4.04
	$p^2\underline{p}\ p\ d$	27.164	9.028	7.642	11.88	3.79
	$p^2p\ p\ \underline{d}$	12.940	7.864	4.437	4.07	1.16
	$p^2\underline{p}\ d\ d$	30.281	7.432	7.799	14.68	4.73
	$p^2p\ \underline{d}\ d$	16.086	6.122	4.506	7.07	2.17
	$\underline{p}\ p\ d^2d$	33.648	8.015	8.005	17.64	5.72
	$p\ p\ d^2\underline{d}$	19.483	6.731	4.623	10.24	3.23
	$p^2\underline{d}\ d\ d$	19.407	4.334	4.617	10.17	3.21
	$\underline{p}\ d^2d\ d$	37.116	6.363	8.249	20.62	6.72
	$p\ d^2\underline{d}\ d$	22.980	4.932	4.778	13.42	4.30
	$d^2\underline{d}\ d\ d$	26.651	3.087	4.976	16.70	5.40
	$di^2\underline{di}\ p_\pi p_\pi$	23.861	11.106	6.323	11.21	3.56
	$di^2di\ \underline{p}_\pi p_\pi$	16.589	9.169	4.978	6.63	2.02
	$\underline{di}\ di\ p_\pi^2 p_\pi$	22.807	11.118	6.021	10.77	3.41
	$di\ di\ p_\pi^2\underline{p}_\pi$	16.848	9.872	4.952	6.94	2.13
	$tr^2\underline{tr}\ tr\ p_\pi$	21.527	10.526	5.938	9.65	3.04
	$tr^2tr\ tr\ \underline{p}_\pi$	16.572	9.407	4.933	6.71	2.05
	$\underline{tr}\ tr\ tr\ p_\pi^2$	20.916	10.683	5.727	9.46	2.97
	$te^2\underline{te}\ te\ te$	20.247	10.299	5.679	8.89	2.78
	$te^2\underline{te}\ te\ d_\pi$	26.388	8.993	6.978	12.43	3.97
	$te^2te\ te\ \underline{d}_\pi$	8.065	6.760	2.774	2.52	0.64
	$\underline{te}\ te\ te\ d_\pi^2$	29.276	9.652	7.003	15.27	4.92
	$pl^2\underline{pl}\ pl\ pl$	21.633	8.503	5.500	10.63	3.37
	$pl^2\underline{pl}\ pl\ p_\pi$	22.331	9.075	5.981	10.37	3.28
	$pl^2pl\ pl\ \underline{p}_\pi$	23.019	8.768	6.226	10.57	3.34
	$\underline{pl}\ pl\ pl\ p_\pi^2$	20.972	9.147	5.739	9.49	2.98
	$tp^2\underline{tp}\ tp\ tp$	23.172	6.333	5.136	12.90	4.13
	$pe^2\underline{pe}\ pe\ pe$	22.006	8.942	5.863	10.28	3.25
	$qp^2\underline{qp}\ qp\ qp$	22.353	7.437	5.290	11.77	3.75
	$oh^2\underline{oh}\ oh\ oh$	23.193	7.996	5.940	11.31	3.59
	$\underline{s}\ p\ p\ p\ d$	29.407	11.166	6.189	17.03	5.52
	$s\ \underline{p}\ p\ p\ d$	21.825	8.486	5.814	10.20	3.22
	$s\ p\ p\ p\ \underline{d}$	6.760	6.404	2.275	2.21	0.54
	$\underline{s}\ p\ p\ d\ d$	32.841	9.848	6.424	19.99	6.51
	$s\ \underline{p}\ p\ d\ d$	26.100	7.535	6.383	13.33	4.27
	$s\ p\ p\ \underline{d}\ d$	11.064	5.341	2.756	5.55	1.66
	$\underline{s}\ p\ d\ d\ d$	36.451	8.500	6.703	23.05	7.54
	$s\ \underline{p}\ d\ d\ d$	30.550	6.556	6.996	16.56	5.36
	$s\ p\ \underline{d}\ d\ d$	15.544	4.250	3.279	8.99	2.81
	$\underline{s}\ d\ d\ d\ d$	40.236	7.124	7.024	26.19	8.59
	$s\ \underline{d}\ d\ d\ d$	20.199	3.130	3.846	12.51	4.00
	$\underline{p}\ p\ p\ d\ d$	30.244	8.482	7.815	14.61	4.70
	$p\ p\ p\ \underline{d}\ d$	16.049	6.655	4.522	7.01	2.15
	$\underline{p}\ p\ d\ d\ d$	33.536	7.425	8.016	17.50	5.67
	$p\ p\ \underline{d}\ d\ d$	19.370	5.487	4.633	10.10	3.19

Table 1 (continued)

Atom	Valence state	b^0	b^1	c	χ_M	χ_P
P	$\mathrm{\underline{p}\ d\ d\ d\ d}$	37.003	6.340	8.259	20.48	6.68
	$\mathrm{p\ \underline{d}\ d\ d\ d}$	22.867	4.290	4.788	13.29	4.26
	$\mathrm{d\ d\ d\ d\ d}$	26.539	3.064	4.986	16.57	5.36
	$\mathrm{\underline{te}\ te\ te\ te\ d_\pi}$	23.709	9.215	6.005	11.70	3.72
	$\mathrm{te\ te\ te\ te\ \underline{d}_\pi}$	5.385	6.372	1.800	1.79	0.39
	$\mathrm{\underline{pl}\ pl\ pl\ pl\ p_\pi}$	20.240	8.677	5.269	9.70	3.05
	$\mathrm{pl\ pl\ pl\ pl\ \underline{p}_\pi}$	20.928	8.428	5.513	9.90	3.12
	$\mathrm{\underline{pe}\ pe\ pe\ pe\ pe}$	20.203	8.661	5.242	9.72	3.06
	$\mathrm{\underline{qp}\ qp\ qp\ qp\ qp}$	20.857	7.178	4.802	11.25	3.57
	$\mathrm{\underline{oh}\ oh\ oh\ oh\ oh}$	21.738	7.747	5.484	10.77	3.41
S	$\mathrm{s^2p^2\underline{p}\ p}$	17.400	9.655	5.003	7.39	2.28
	$\mathrm{\underline{s}\ p^2p^2p}$	24.357	12.311	4.271	15.81	5.11
	$\mathrm{s\ p^2p^2\underline{p}}$	18.228	12.408	4.905	8.42	2.62
	$\mathrm{s^2p^2\underline{p}\ d}$	20.710	8.955	5.820	9.07	2.84
	$\mathrm{s^2p^2p\ \underline{d}}$	5.395	7.980	3.680	−1.96	−0.87
	$\mathrm{s^2\underline{p}\ p\ d^2}$	24.072	13.415	6.757	10.56	3.34
	$\mathrm{\underline{s}\ p^2p^2d}$	28.096	10.945	5.197	17.70	5.74
	$\mathrm{s\ p^2p^2\underline{d}}$	5.582	10.527	3.037	−0.49	−0.37
	$\mathrm{\underline{s}\ p^2p\ d^2}$	31.864	14.140	6.233	19.40	6.31
	$\mathrm{s\ p^2\underline{p}\ d^2}$	24.665	14.697	6.214	12.24	3.91
	$\mathrm{s^2p^2\underline{d}\ d}$	9.206	7.165	4.690	−0.17	−0.27
	$\mathrm{s^2\underline{p}\ d^2d}$	26.933	13.065	7.550	11.83	3.77
	$\mathrm{s^2p\ d^2\underline{d}}$	13.032	12.141	5.794	1.44	0.28
	$\mathrm{\underline{s}\ p^2d^2d}$	35.131	13.135	7.126	20.88	6.81
	$\mathrm{s\ p^2d^2\underline{d}}$	12.970	13.216	4.703	3.56	0.99
	$\mathrm{\underline{s}\ p\ d^2d^2}$	38.414	17.047	8.112	22.19	7.25
	$\mathrm{s\ \underline{p}\ d^2d^2}$	30.154	18.052	7.445	15.26	4.92
	$\mathrm{s^2d^2\underline{d}\ d}$	16.358	11.660	6.754	2.85	0.75
	$\mathrm{\underline{s}\ d^2d^2d}$	41.197	16.375	8.955	23.29	7.62
	$\mathrm{s\ d^2d^2\underline{d}}$	19.388	16.954	6.268	6.85	2.10
	$\mathrm{p^2p^2\underline{p}\ d}$	23.819	11.202	6.884	10.05	3.1
	$\mathrm{p^2p^2p\ \underline{d}}$	7.442	10.676	4.096	−0.75	−0.46
	$\mathrm{p^2\underline{p}\ p\ d^2}$	26.931	14.190	7.372	12.19	3.89
	$\mathrm{p^2p^2\underline{d}\ d}$	11.003	8.765	4.658	1.69	0.36
	$\mathrm{p^2\underline{p}\ d^2d}$	29.543	12.745	7.717	14.11	4.53
	$\mathrm{p^2p\ d^2\underline{d}}$	14.580	12.269	5.313	3.95	1.12
	$\mathrm{\underline{p}\ p\ d^2d^2}$	32.171	16.449	8.155	15.86	5.12
	$\mathrm{p^2d^2\underline{d}\ d}$	17.657	10.692	5.825	6.01	1.81
	$\mathrm{\underline{p}\ d^2d^2d}$	34.298	15.338	8.450	17.40	5.64
	$\mathrm{p\ d^2d^2\underline{d}}$	20.749	14.912	6.430	7.89	2.44
	$\mathrm{d^2d^2\underline{d}\ d}$	23.341	13.669	6.891	9.56	3.01
	$\mathrm{di^2di^2\underline{p}_\pi p_\pi}$	17.400	9.655	5.003	7.39	2.28
	$\mathrm{di^2\underline{di}\ p_\pi^2p_\pi}$	23.191	11.529	5.408	12.37	3.95
	$\mathrm{di^2di\ p_\pi^2\underline{p}_\pi}$	17.815	11.522	4.954	7.91	2.45
	$\mathrm{\underline{di}\ di\ p_\pi^2p_\pi^2}$	22.726	11.644	5.305	12.12	3.86
	$\mathrm{tr^2tr^2\underline{tr}\ p_\pi}$	21.775	11.026	5.445	10.89	3.45
	$\mathrm{tr^2tr^2tr\ \underline{p}_\pi}$	17.678	11.009	4.972	7.73	2.39
	$\mathrm{tr^2\underline{tr}\ tr\ p_\pi^2}$	21.662	11.601	5.389	10.88	3.45

Table 1 (continued)

Atom	Valence state	b^0	b^1	c	χ_M	χ_P
S	$te^2te^2\underline{te}\ te$	20.861	11.347	5.361	10.14	3.20
	$te^2te^2\underline{te}\ d_\pi$	24.762	10.550	6.541	11.68	3.72
	$te^2te^2te\ \underline{d_\pi}$	6.033	9.808	3.665	–1.30	–0.64
	$te^2te^2\underline{d_\pi}d_\pi$	10.332	8.470	4.774	0.78	0.06
	$te^2\underline{te}\ te\ d_\pi^2$	28.035	14.340	7.218	13.60	4.36
	$te^2\underline{te}\ d_\pi^2d_\pi$	31.304	13.372	8.139	15.03	4.84
	$te^2te\ \underline{d_\pi^2}d_\pi$	13.973	12.914	5.356	3.26	0.89
	$\underline{te}\ te\ d_\pi^2d_\pi^2$	33.916	16.851	8.533	16.85	5.45
	$pl^2pl^2\underline{pl}\ pl$	22.007	11.907	6.036	9.93	3.13
	$pl^2pl^2\underline{pl}\ p_\pi$	21.386	12.170	5.999	9.39	2.95
	$pl^2pl^2pl\ \underline{p_\pi}$	22.536	12.453	6.099	10.34	3.27
	$pl^2\underline{pl}\ pl\ p_\pi^2$	20.617	11.641	5.748	9.12	2.86
	$tp^2tp^2\underline{tp}\ tp$	22.982	12.725	6.573	9.84	3.10
	$pe^2pe^2\underline{pe}\ pe$	21.440	12.007	5.924	9.59	3.02
	$qp^2qp^2\underline{qp}\ qp$	21.519	12.470	6.072	9.37	2.94
	$oh^2oh^2\underline{oh}\ oh$	21.906	12.437	6.276	9.35	2.94
	$s^2\underline{p}\ p\ p\ d$	20.887	10.683	5.992	8.90	2.78
	$s^2p\ p\ p\ \underline{d}$	5.572	8.107	3.851	–2.13	–0.92
	$\underline{s}\ p^2p\ p\ d$	28.273	12.470	5.369	17.54	5.69
	$s\ p^2\underline{p}\ p\ d$	21.605	11.111	5.673	10.26	3.24
	$s\ p^2p\ p\ \underline{d}$	5.759	8.893	3.209	–0.66	–0.43
	$\underline{s}\ p\ p\ p\ d^2$	32.040	14.268	6.404	19.23	6.26
	$s\ \underline{p}\ p\ p\ d^2$	24.842	13.266	6.385	12.07	3.85
	$s^2\underline{p}\ p\ d\ d$	23.990	11.799	6.810	10.37	3.28
	$s^2p\ p\ \underline{d}\ d$	9.382	9.109	4.862	–0.34	–0.32
	$\underline{s}\ p^2p\ d\ d$	31.782	12.728	6.287	19.21	6.25
	$s\ p^2\underline{p}\ d\ d$	24.584	11.585	6.268	12.05	3.84
	$s\ p^2p\ \underline{d}\ d$	9.445	9.252	3.995	1.45	0.28
	$\underline{s}\ p\ p\ d^2d$	35.308	14.788	7.298	20.71	6.75
	$s\ \underline{p}\ p\ d^2d$	27.578	14.003	6.955	13.67	4.39
	$s\ p\ p\ d^2\underline{d}$	13.146	11.707	4.874	3.40	0.94
	$s^2\underline{p}\ d\ d\ d$	26.852	13.114	7.604	11.64	3.71
	$s^2p\ \underline{d}\ d\ d$	12.950	10.309	5.847	1.26	0.22
	$\underline{s}\ p^2d\ d\ d$	35.050	13.184	7.180	20.69	6.75
	$s\ p^2\underline{d}\ d\ d$	12.888	9.811	4.757	3.38	0.93
	$\underline{s}\ p\ d^2d\ d$	38.333	15.507	8.166	22.00	7.19
	$s\ \underline{p}\ d^2d\ d$	30.073	14.938	7.499	15.07	4.86
	$s\ p\ d^2\underline{d}\ d$	16.348	12.528	5.610	5.13	1.52
	$s^2\underline{d}\ d\ d\ d$	16.277	11.709	6.808	2.66	0.69
	$\underline{s}\ d^2d\ d\ d$	41.116	16.424	9.008	23.10	7.55
	$s\ d^2\underline{d}\ d\ d$	19.306	13.548	6.322	6.66	2.03
	$p^2\underline{p}\ p\ d\ d$	26.850	12.700	7.426	12.00	3.82
	$p^2p\ p\ \underline{d}\ d$	11.180	10.584	4.830	1.52	0.30
	$\underline{p}\ p\ p\ d^2d$	29.720	14.475	7.889	13.94	4.48
	$p\ p\ p\ d^2\underline{d}$	14.757	12.396	5.484	3.79	1.07
	$p^2\underline{p}\ d\ d\ d$	29.462	12.794	7.771	13.92	4.47
	$p^2p\ \underline{d}\ d\ d$	14.499	10.563	5.367	3.77	1.06
	$\underline{p}\ p\ d^2d\ d$	32.089	14.831	8.209	15.67	5.06

Table 1 (continued)

Atom	Valence state	b^0	b^1	c	χ_M	χ_P
S	$p\ p\ d^2\underline{d}\ d$	17.834	12.638	5.996	5.84	1.76
	$p^2\underline{d}\ d\ d\ d$	17.576	10.741	5.879	5.82	1.75
	$\underline{p}\ d^2d\ d\ d$	34.217	15.386	8.503	17.21	5.58
	$p\ d^2\underline{d}\ d\ d$	20.668	13.078	6.483	7.70	2.38
	$d^2\underline{d}\ d\ d\ d$	23.260	13.718	6.945	9.37	2.94
	$te^2\underline{te}\ te\ te\ d_\pi$	24.350	11.433	6.214	11.92	3.80
	$te^2te\ te\ te\ \underline{d}_\pi$	5.621	8.760	3.338	−1.05	−0.56
	$te^2\underline{te}\ te\ d_\pi d_\pi$	27.949	12.481	7.277	13.40	4.29
	$te^2te\ te\ \underline{d}_\pi d_\pi$	9.919	9.758	4.447	1.03	0.14
	$\underline{te}\ te\ te\ te\ d_\pi^2$	27.623	13.292	6.891	13.84	4.44
	$\underline{te}\ te\ te\ d_\pi^2 d_\pi$	30.891	14.215	7.812	15.27	4.92
	$te\ te\ te\ d_\pi^2\underline{d}_\pi$	13.561	11.866	5.029	3.50	0.97
	$pl^2\underline{pl}\ pl\ pl\ p_\pi$	21.364	11.342	5.902	9.56	3.01
	$pl^2pl\ pl\ pl\ \underline{p}_\pi$	22.514	11.558	6.003	10.51	3.32
	$\underline{pl}\ pl\ pl\ pl\ p_\pi^2$	20.595	10.746	5.652	9.29	2.92
	$pe^2\underline{pe}\ pe\ pe\ pe$	21.379	11.236	5.817	9.75	3.07
	$qp^2\underline{qp}\ qp\ qp\ qp$	21.627	11.834	6.005	9.62	3.02
	$oh^2\underline{oh}\ oh\ oh\ oh$	22.104	11.800	6.335	9.43	2.96
	$\underline{s}\ p\ p\ p\ d\ d$	31.959	13.440	6.458	19.04	6.19
	$s\ \underline{p}\ p\ p\ d\ d$	24.760	12.042	6.439	11.88	3.79
	$s\ p\ p\ p\ \underline{d}\ d$	9.621	9.391	4.167	1.29	0.23
	$\underline{s}\ p\ p\ d\ d\ d$	35.226	14.226	7.351	20.52	6.69
	$s\ \underline{p}\ p\ d\ d\ d$	27.497	13.086	7.008	13.48	4.32
	$s\ p\ p\ \underline{d}\ d\ d$	13.065	10.338	4.928	3.21	0.87
	$\underline{s}\ p\ d\ d\ d\ d$	38.252	15.237	8.219	21.81	7.12
	$s\ \underline{p}\ d\ d\ d\ d$	29.991	14.354	7.553	14.89	4.80
	$s\ p\ \underline{d}\ d\ d\ d$	16.266	11.510	5.664	4.94	1.45
	$\underline{s}\ d\ d\ d\ d\ d$	41.035	16.473	9.062	22.91	7.49
	$s\ \underline{d}\ d\ d\ d\ d$	19.225	12.833	6.375	6.47	1.97
	$\underline{p}\ p\ p\ d\ d\ d$	29.639	13.883	7.942	13.75	4.41
	$p\ p\ p\ \underline{d}\ d\ d$	14.676	11.392	5.538	3.60	1.00
	$\underline{p}\ p\ d\ d\ d\ d$	32.008	14.547	8.262	15.48	5.00
	$p\ p\ \underline{d}\ d\ d\ d$	17.752	11.959	6.050	5.65	1.69
	$\underline{p}\ d\ d\ d\ d\ d$	34.136	15.435	8.557	17.02	5.51
	$p\ \underline{d}\ d\ d\ d\ d$	20.587	12.751	6.537	7.51	2.32
	$\underline{te}\ te\ te\ te\ d_\pi d_\pi$	27.537	12.198	6.949	13.64	4.38
	$te\ te\ te\ te\ \underline{d}_\pi d_\pi$	9.507	9.177	4.120	1.27	0.22
	$\underline{oh}\ oh\ oh\ oh\ oh\ oh$	22.301	11.163	6.393	9.52	2.99
Cl	$s^2p^2p^2\underline{p}$	20.800		5.716	9.37	2.94
	$\underline{s}\ p^2p^2p^2$	28.914		4.846	19.22	6.25
K	s	5.534		1.193	3.15	0.85
	p	3.452		0.725	2.00	0.47
Ca	$\underline{s}\ p$	9.519	4.633	2.428	4.66	1.36
	$s\ \underline{p}$	6.057	4.329	2.096	1.87	0.42
	$p\ p$	6.211	3.948	1.831	2.55	0.65
	$di\ di$	8.146	4.120	2.380	3.39	0.93
	$\underline{di}\ p_\pi$	8.383	4.598	2.648	3.09	0.83
	$di\ \underline{p}_\pi$	6.440	4.418	2.270	1.90	0.43

Table 1 (continued)

Atom	Valence state	b^0	b^1	c	χ_M	χ_P
Ca	tr tr	7.862	4.477	2.611	2.64	0.68
	$\underline{tr}\ p_\pi$	7.987	4.548	2.703	2.58	0.66
	$tr\ \underline{p}_\pi$	6.532	4.535	2.291	1.95	0.45
	te te	7.535	4.451	2.522	2.49	0.63
Ga	$s^2\underline{p}$	9.138		3.140	2.86	0.75
	$\underline{s}\ p^2$	18.022		3.869	10.29	3.25
	$p^2\underline{p}$	6.618		-0.876	8.37	2.61
	$di^2\underline{di}$	15.603		4.167	7.27	2.24
	$di^2\underline{p}_\pi$	7.659		0.607	6.45	1.96
	$\underline{di}\ p_\pi^2$	12.186		1.362	9.46	2.97
	$tr^2\underline{tr}$	12.025		2.056	7.91	2.45
	$tr^2\underline{p}_\pi$	7.287		0.006	7.28	2.24
	$\underline{tr}\ p_\pi^2$	10.300		0.586	9.13	2.86
	$te^2\underline{te}$	10.444		1.158	8.13	2.52
	$\underline{s}\ p\ p$	19.036		4.450	10.14	3.20
	$s\ \underline{p}\ p$	9.173		2.424	4.33	1.25
	p p p	7.631		-0.295	8.22	2.56
	$\underline{di}\ di\ p_\pi$	15.158		3.964	7.23	2.22
	$di\ di\ \underline{p}_\pi$	9.248		2.669	3.91	1.11
	$\underline{di}\ p_\pi p_\pi$	13.199		1.943	9.31	2.92
	$di\ \underline{p}_\pi p_\pi$	8.401		1.064	6.27	1.90
	tr tr tr	13.429		3.676	6.08	1.84
	$\underline{tr}\ tr\ p_\pi$	12.354		2.367	7.62	2.35
	$tr\ tr\ \underline{p}_\pi$	8.691		1.626	5.44	1.62
	te te te	11.764		2.541	6.68	2.04
Ge	$s^2\underline{p}\ p$	10.564		3.221	4.12	1.18
	$\underline{s}\ p^2p$	24.962		5.796	13.37	4.29
	$s\ p^2\underline{p}$	12.541		2.520	7.50	2.31
	$p^2\underline{p}\ p$	20.526		4.492	11.54	3.67
	$di^2\underline{di}\ p_\pi$	20.100		5.287	9.53	2.99
	$di^2di\ \underline{p}_\pi$	11.663		2.908	5.85	1.76
	$\underline{di}\ di\ p_\pi^2$	19.169		4.368	10.43	3.30
	$di^2\underline{p}_\pi p_\pi$	17.140		4.311	8.52	2.66
	$\underline{di}\ p_\pi^2p_\pi$	22.837		5.237	12.36	3.95
	$di\ p_\pi^2\underline{p}_\pi$	16.533		3.507	9.52	2.99
	$tr^2\underline{tr}\ tr$	17.272		4.294	8.69	2.71
	$tr^2\underline{tr}\ p_\pi$	20.448		5.084	10.28	3.25
	$tr^2tr\ \underline{p}_\pi$	15.278		3.622	8.03	2.49
	$\underline{tr}\ tr\ p_\pi^2$	19.570		4.404	10.76	3.41
	$te^2\underline{te}\ te$	18.258		4.391	9.48	2.98
	$\underline{s}\ p\ p\ p$	24.448		5.865	12.72	4.07
	$s\ \underline{p}\ p\ p$	12.027		2.590	6.85	2.09
	$\underline{di}\ di\ p_\pi p_\pi$	18.656		4.437	9.78	3.08
	$di\ di\ \underline{p}_\pi p_\pi$	11.284		2.381	6.52	1.98
	$\underline{tr}\ tr\ tr\ p_\pi$	16.207		3.774	8.66	2.70
	$tr\ tr\ tr\ \underline{p}_\pi$	11.034		2.310	6.41	1.95
	te te te te	14.887		3.408	8.07	2.51
As	$s^2p^2\underline{p}$	14.365		4.345	5.68	1.70

Table 1 (continued)

Atom	Valence state	b^0	b^1	c	χ_M	χ_P
As	$\underline{s}\ p^2p^2$	21.354		4.476	12.40	3.96
	$p^2p^2\underline{p}$	16.614		3.295	10.02	3.16
	$di^2di^2\underline{p}_\pi$	14.365		4.345	5.68	1.70
	$di^2\underline{di}\ p_\pi^2$	18.766		4.713	9.34	2.93
	$di^2p_\pi^2\underline{p}_\pi$	15.220		3.639	7.94	2.46
	$\underline{di}\ p_\pi^2p_\pi^2$	19.194		4.095	11.00	3.49
	$tr^2tr^2\underline{tr}$	17.503		4.658	8.19	2.54
	$tr^2tr^2\underline{p}_\pi$	14.875		3.834	7.21	2.21
	$tr^2\underline{tr}\ p_\pi^2$	17.821		4.187	9.45	2.97
	$te^2te^2\underline{te}$	17.047		4.229	8.59	2.68
	$s^2\underline{p}\ p\ p$	13.306	8.802	3.942	5.42	1.62
	$\underline{s}\ p^2p\ p$	20.296	13.016	4.074	12.15	3.88
	$s\ p^2\underline{p}\ p$	16.468	11.405	4.311	7.85	2.43
	$di^2\underline{di}\ p_\pi p_\pi$	17.708	11.522	4.310	9.09	2.85
	$di^2di\ \underline{p}_\pi p_\pi$	14.886	10.462	4.126	6.63	2.02
	$\underline{di}\ di\ p_\pi^2p_\pi$	19.077	12.060	4.540	10.00	3.15
	$di\ di\ p_\pi^2\underline{p}_\pi$	16.602	11.405	4.401	7.80	2.41
	$tr^2\underline{tr}\ tr\ p_\pi$	17.401	11.426	4.397	8.61	2.69
	$tr^2tr\ tr\ \underline{p}_\pi$	15.473	10.842	4.228	7.02	2.15
	$\underline{tr}\ tr\ tr\ p_\pi^2$	18.421	11.808	4.580	9.26	2.90
	$te^2\underline{te}\ te\ te$	17.242	11.399	4.431	8.38	2.61
Se	$s^2p^2\underline{p}\ p$	16.343	8.957	4.666	7.01	2.15
	$\underline{s}\ p^2p^2p$	25.659	12.559	5.161	15.34	4.95
	$s\ p^2p^2\underline{p}$	20.731	10.928	6.291	8.15	2.53
	$di^2di^2\underline{p}_\pi p_\pi$	16.349	8.957	4.669	7.01	2.15
	$di^2\underline{di}\ p_\pi^2p_\pi$	22.805	11.060	5.516	11.77	3.75
	$di^2di\ p_\pi^2\underline{p}_\pi$	18.539	10.112	5.479	7.58	2.34
	$\underline{di}\ di\ p_\pi^2p_\pi^2$	24.135	11.275	6.196	11.74	3.74
	$tr^2\underline{tr}^2tr\ p_\pi$	21.055	10.426	5.367	10.32	3.26
	$tr^2tr^2tr\ \underline{p}_\pi$	17.810	9.765	5.209	7.39	2.28
	$tr^2\underline{tr}\ tr\ p_\pi^2$	22.131	10.816	5.849	10.43	3.30
	$te^2te^2\underline{te}\ te$	20.908	10.469	5.615	9.68	3.05
Br	$s^2p^2p^2\underline{p}$	17.904		4.796	8.31	2.59
	$\underline{s}\ p^2p^2p^2$	25.964		3.883	18.20	5.91
Rb	s	6.023		1.846	2.33	0.58
	p	2.733		0.135	2.46	0.62
Sr	$\underline{s}\ p$	8.854	4.455	2.229	4.40	1.27
	$s\ \underline{p}$	5.681	3.966	2.026	1.63	0.34
	p p	6.777	4.456	2.574	1.63	0.34
	di di	7.552	3.862	2.201	3.15	0.85
	$\underline{di}\ p_\pi$	8.039	4.453	2.625	2.79	0.73
	$di\ \underline{p}_\pi$	6.227	4.223	2.298	1.63	0.34
	tr tr	7.472	4.284	2.550	2.37	0.59
	$\underline{tr}\ p_\pi$	7.739	4.398	2.729	2.28	0.56
	$tr\ \underline{p}_\pi$	6.354	4.374	2.334	1.69	0.36
	te te	6.926	3.971	2.201	2.52	0.64
In	$s^2\underline{p}$	8.903		3.117	2.67	0.69
	$\underline{s}\ p^2$	16.196		3.348	9.50	2.99

Table 1 (continued)

Atom	Valence state	b^0	b^1	c	χ_M	χ_P
In	$\mathrm{p^2\underline{p}}$	8.700		1.836	5.03	1.48
	$\mathrm{di^2\underline{di}}$	13.960		3.672	6.62	2.02
	$\mathrm{di^2\underline{p}_\pi}$	8.779		2.366	4.05	1.15
	$\mathrm{\underline{di}\ p_\pi^2}$	12.782		2.925	6.93	2.12
	$\mathrm{tr^2\underline{tr}}$	12.073		3.082	5.91	1.78
	$\mathrm{tr^2\underline{p}_\pi}$	8.790		2.186	4.42	1.28
	$\mathrm{\underline{tr}\ p_\pi^2}$	11.494		2.635	6.22	1.88
	$\mathrm{te^2\underline{te}}$	11.178		2.778	5.62	1.68
	$\mathrm{\underline{s}\ p\ p}$	15.933		3.337	9.26	2.90
	$\mathrm{s\ \underline{p}\ p}$	8.981		2.790	3.40	0.94
	$\mathrm{p\ p\ p}$	8.366		1.754	4.86	1.43
	$\mathrm{\underline{di}\ di\ p_\pi}$	13.304		3.469	6.37	1.93
	$\mathrm{di\ di\ \underline{p}_\pi}$	8.900		2.799	3.30	0.90
	$\mathrm{\underline{di}\ p_\pi p_\pi}$	12.446		2.842	6.76	2.07
	$\mathrm{di\ \underline{p}_\pi p_\pi}$	8.674		2.272	4.13	1.18
	$\mathrm{tr\ tr\ tr}$	12.022		3.344	5.33	1.59
	$\mathrm{\underline{tr}\ tr\ p_\pi}$	11.608		2.946	5.72	1.71
	$\mathrm{tr\ tr\ \underline{p}_\pi}$	8.739		2.448	3.84	1.08
	$\mathrm{te\ te\ te}$	11.052		2.956	5.14	1.52
Sn	$\mathrm{s^2\underline{p}\ p}$	9.860		2.915	4.03	1.15
	$\mathrm{\underline{s}\ p^2p}$	20.426		4.081	12.26	3.91
	$\mathrm{s\ p^2\underline{p}}$	9.874		1.361	7.15	2.20
	$\mathrm{p^2\underline{p}\ p}$	12.981		0.876	11.23	3.57
	$\mathrm{di^2\underline{di}\ p_\pi}$	17.236		4.195	8.85	2.77
	$\mathrm{di^2di\ \underline{p}_\pi}$	10.131		2.226	5.68	1.70
	$\mathrm{\underline{di}\ di\ p_\pi^2}$	15.920		3.106	9.71	3.06
	$\mathrm{di^2\underline{p}_\pi p_\pi}$	12.031		1.878	8.27	2.57
	$\mathrm{\underline{di}\ p_\pi^2p_\pi}$	16.756		2.531	11.69	3.72
	$\mathrm{di\ p_\pi^2\underline{p}_\pi}$	11.428		1.119	9.19	2.88
	$\mathrm{tr^2\underline{tr}\ tr}$	14.718		3.286	8.15	2.53
	$\mathrm{tr^2\underline{tr}\ p_\pi}$	15.589		2.930	9.73	3.06
	$\mathrm{tr^2tr\ \underline{p}_\pi}$	11.235		1.729	7.78	2.41
	$\mathrm{\underline{tr}\ tr\ p_\pi^2}$	14.793		2.301	10.19	3.22
	$\mathrm{te^2\underline{te}\ te}$	14.166		2.587	8.99	2.81
	$\mathrm{\underline{s}\ p\ p\ p}$	20.257		4.095	12.07	3.85
	$\mathrm{s\ \underline{p}\ p\ p}$	9.705		1.375	6.96	2.13
	$\mathrm{\underline{di}\ di\ p_\pi p_\pi}$	15.749		3.119	9.51	2.99
	$\mathrm{di\ di\ \underline{p}_\pi p_\pi}$	9.532		1.428	6.68	2.04
	$\mathrm{\underline{tr}\ tr\ tr\ p_\pi}$	13.828		2.646	8.54	2.66
	$\mathrm{tr\ tr\ tr\ \underline{p}_\pi}$	9.472		1.444	6.58	2.01
	$\mathrm{te\ te\ te\ te}$	12.788		2.382	8.02	2.49
Sb	$\mathrm{s^2p^2\underline{p}}$	13.487		4.189	5.11	1.51
	$\mathrm{\underline{s}\ p^2p^2}$	25.384		6.043	13.30	4.26
	$\mathrm{p^2p^2\underline{p}}$	21.797		5.282	11.23	3.57
	$\mathrm{di^2di^2\underline{p}_\pi}$	11.375		3.485	4.41	1.27
	$\mathrm{di^2\underline{di}\ p_\pi^2}$	21.689		5.867	9.96	3.14
	$\mathrm{di^2p_\pi^2\underline{p}_\pi}$	18.799		5.050	8.70	2.72
	$\mathrm{\underline{di}\ p_\pi^2p_\pi^2}$	23.814		5.886	12.04	3.84

Table 1 (continued)

Atom	Valence state	b^0	b^1	c	χ_M	χ_P
Sb	$tr^2tr^2\underline{tr}$	19.457		5.475	8.51	2.65
	$tr^2tr^2\underline{p}_\pi$	17.285		4.832	7.62	2.35
	$tr^2\underline{tr}\ p_\pi^2$	21.837		5.743	10.35	3.27
	$te^2te^2\underline{te}$	20.300		5.515	9.27	2.91
	$s^2\underline{p}\ p\ p$	12.563	6.894	3.810	4.94	1.45
	$\underline{s}\ p^2p\ p$	24.461	9.803	5.665	13.13	4.21
	$s\ p^2\underline{p}\ p$	15.729	7.672	4.052	7.62	2.36
	$di^2\underline{di}\ p_\pi p_\pi$	20.765	8.864	5.489	9.79	3.08
	$di^2di\ \underline{p}_\pi p_\pi$	14.146	7.398	3.931	6.28	1.90
	$\underline{di}\ di\ p_\pi^2p_\pi$	20.755	8.704	5.188	10.38	3.28
	$di\ di\ p_\pi^2\underline{p}_\pi$	15.150	7.671	3.894	7.36	2.27
	$tr^2\underline{tr}\ tr\ p_\pi$	18.879	8.358	4.989	8.90	2.78
	$tr^2tr\ tr\ \underline{p}_\pi$	14.416	7.484	3.901	6.61	2.02
	$\underline{tr}\ tr\ tr\ p_\pi^2$	18.969	8.353	4.813	9.34	2.93
	$te^2\underline{te}\ te\ te$	17.869	8.146	4.708	8.45	2.63
Te	$s^2p^2\underline{p}\ p$	15.433	7.548	4.395	6.64	2.03
	$\underline{s}\ p^2p^2p$	26.777	9.584	6.005	14.77	4.76
	$s\ p^2p^2\underline{p}$	20.894	8.119	6.095	8.70	2.72
	$di^2di^2\underline{p}_\pi p_\pi$	15.433	7.548	4.395	6.64	2.03
	$di^2\underline{di}\ p_\pi^2p_\pi$	22.928	8.849	5.808	11.31	3.59
	$di^2di\ p_\pi^2\underline{p}_\pi$	18.164	7.870	5.246	7.67	2.37
	$\underline{di}\ di\ p_\pi^2p_\pi^2$	24.641	8.490	6.454	11.73	3.74
	$tr^2tr^2\underline{tr}\ p_\pi$	20.836	8.472	5.473	9.89	3.12
	$tr^2tr^2tr\ \underline{p}_\pi$	17.255	7.771	4.963	7.33	2.26
	$tr^2\underline{tr}\ tr\ p_\pi^2$	22.203	8.392	5.948	10.31	3.26
	$te^2te^2\underline{te}\ te$	20.746	8.262	5.627	9.49	2.98
I	$s^2p^2p^2\underline{p}$	17.328		4.651	8.03	2.49
	$\underline{s}\ p^2p^2p^2$	20.397		2.390	15.62	5.04

[a] The values for the underlined orbitals are displayed

The symbols for the valence states as defined by Mulliken[12)] were extended by the following:

Symbol	Geometry	Hybridisation	Point-Group
pl	square planar	dsp^2	D_{4h}
tp	trigonal pyramidal	d^2sp	C_{3v}
pe	pentagonal bipyramidal	dsp^3	D_{3h}
qp	quadratic pyramidal	d^2sp^2	C_{4v}
oh	octahedral	d^2sp^3	O_h

Dimensions are: [V] for b^0 and χ_M, [V/e] for b^1 and c, [$\sqrt{eV}$] for χ_P

Table 2. Ab-initio partial charges[25] compared with our values[a]

Compound	STO-3G//	3-21G//	4-31G//	6-31G//	from χ's
$\underline{H}F$	209	453	479	517	164
$\underline{H}_2O$	186	370	400	438	173
$N\underline{H}_3$	160	285	305	336	81
$C\underline{H}_4$	64	197	153	165	15
$\underline{C}H_4$	-255	-789	-610	-659	-60
$C\underline{H}{\equiv}CH$	109	336	299	279	123
$\underline{C}H{\equiv}CH$	-109	-336	-299	-279	-123
$C\underline{H}_2{=}CH_2$	62	213	165	178	45
$\underline{C}H_2{=}CH_2$	-124	-425	-331	-356	-91
$\underline{H}CN$	149	371	330	316	214
$H\underline{C}N$	9	55	13	65	8
$C\underline{H}_3F$	63	187	159	153	65
$\underline{C}H_3F$	-32	-146	-21	-64	9
$CH_3O\underline{H}$	189	374	397	439	143
$C\underline{H}_3OH$	62	199	157	156	72
$\underline{C}H_3OH$	-66	-259	-126	-164	18
$CH_3N\underline{H}_2$	157	290	307	336	73
$C\underline{H}_3NH_2$	61	193	153	156	43
$\underline{C}H_3NH_2$	-89	-381	-260	-290	-22
$C\underline{H}_3CN$	100	271	222	229	88
$\underline{C}H_3CN$	-166	-613	-443	-533	39
$CH_3\underline{C}N$	67	312	195	308	-41
$\underline{H}_2CO$	63	183	161	155	144
$H_2\underline{C}O$	59	117	169	130	45
$HCO_2\underline{H}$	240	432	455	483	226
$\underline{H}CO_2H$	88	251	231	211	208
$H\underline{C}O_2H$	257	623	628	542	138
$\underline{H}CF_3$	69	223	219	175	181
$H\underline{C}F_3$	376	941	991	921	166
Std. Dev.:	70	80	72	77	-
Corr. Coeff.:	0.760	0.720	0.747	0.718	-

[a] The charge [in 10^{-3} e] of the underlined atom is displayed

E. Group Electronegativities and Hardness

The orbital electronegativity of an orbital which is free to form a bond in a molecular group can be computed quite analogously. Again we will have as many linear equations of type (25) as there are bonds in the molecular group to determine the Δq's in these bonds.

For the free group orbital, i, for which the electronegativity is desired, the charge shift, $\Delta q_i = 0$ will be fixed. With this the charge distribution in the group can be computed to yield the rest charge r_i^0 for the orbital of interest. The group electronegativity is then given immediately as

$$\chi_g = \chi_i(r_i^0, q_i^0) = b_i^0 + b_i^1 r_i^0 + 2\, c_i q_i^0 \tag{26}$$

The effective hardness of the free group orbital will not be just c_i of this free orbital, because a charge shift of Δq_i into this orbital will result in a charge rearrangement in the other orbitals of the group, resulting in a different rest charge $r_i' \neq r_i^0$. The effective hardness of the group orbital can be obtained by calculating the charge distribution in the group once with $\Delta q_i = 0$, yielding the group electronegativity, Eq. (26), and a second time with some finite charge shift, say $\Delta q_i = 0.1$ in the free group orbital. This will yield

$$\chi_g' = \chi_i(r_i', q_i^0 + \Delta q) = b_i^0 + b_i^1 r_i' + 2\, c_i(q_i^0 + \Delta q_i) \tag{27}$$

A finite difference approximation for the derivative will now yield the effective hardness of the free group orbital.

$$\eta_i = \tfrac{1}{2}\frac{\partial^2 E}{\partial q_i^2} = \tfrac{1}{2}\frac{\partial \chi_i}{\partial q_i} = \tfrac{1}{2}\frac{\chi_i(r_i', q_i^0 + \Delta q) - \chi_i(r_i^0, q_i^0)}{\Delta q} \tag{28}$$

In Table 3 we give the effective group electronegativities and hardness parameters for a number of important molecular groups. These results may now be used directly to calculate the charge shift between two groups joined together to form a large molecule. Provided the charge shift is not too large the "back of an envelope" calculation with the effective group-electronegativity parameters will suffice, and there is no need to solve again the set of linear equations for the resulting combined molecule.

Table 3 Group electronegativities and hardness parameters[a]

Group	χ_M	χ_P	$\eta = c_{eff}$	η_{Lit} [b]
–H	7.18	2.20	6.422	6.42
–F	12.14	3.87	8.726	7.01
–Cl	9.37	2.94	5.716	4.70
–Br	8.31	2.59	4.796	4.24
–I	8.03	2.49	4.651	3.70
$-CF_3$	10.62	3.36	3.115	–
$-CCl_3$	8.99	2.81	2.669	–
$-CBr_3$	8.25	2.56	2.509	–
$-CI_3$	8.03	2.49	2.482	–
$-CHF_2$	9.44	2.96	2.991	–
$-CHCl_2$	8.50	2.65	2.706	–
$-CHBr_2$	8.01	2.48	2.590	–
$-CHI_2$	7.86	2.44	2.570	–
$-CH_2F$	8.39	2.61	2.882	–
$-CH_2Cl$	7.99	2.48	2.744	–
$-CH_2Br$	7.75	2.40	2.681	–
$-CH_2I$	7.67	2.37	2.670	–
$-CH_3$	7.45	2.30	2.784	4.0
$-C_2H_5$	7.50	2.31	2.478	–
$-C_3H_7$	7.52	2.32	2.435	–
$-(i\text{-}C_3H_7)$	7.54	2.33	2.265	–
$-C_4H_9$	7.53	2.32	2.429	–

Table 3 (continued)

Group	χ_M	χ_P	$\eta = c_{eff}$	η_{Lit} [b]
$-(n\text{-}C_4H_9)$ (2)	7.55	2.33	2.234	–
$-(t\text{-}C_4H_9)$	8.64	2.69	2.267	–
$-C_5H_{11}$	7.53	2.32	2.428	–
$-(n\text{-}C_5H_{11})$ (2)	7.56	2.33	2.229	–
$-(n\text{-}C_5H_{11})$ (3)	7.57	2.34	2.204	–
$-(i\text{-}C_5H_{11})$	7.53	2.32	2.424	–
$-(i\text{-}C_5H_{11})$ (2)	7.57	2.33	2.210	–
$-(neo\text{-}C_5H_{11})$	7.55	2.33	2.376	–
$-C_6H_{13}$	7.53	2.32	2.427	–
$-(c\text{-}C_6H_{11})$	7.59	2.34	2.203	–
$-(c\text{-}C_5H_9)$	7.59	2.34	2.217	–
$-CH=CH_2$	8.01	2.48	2.786	–
$-CH_2-CH=CH_2$	7.71	2.38	2.479	–
$-CH=CH-CH_3$	7.97	2.47	2.560	–
$-CH_2-CH=CH-CH_3$	7.70	2.38	2.459	–
$-C(CH_3)=CH-CH_3$	7.59	2.34	2.455	–
$-CH=CH-CH_2-CH_3$	7.99	2.48	2.623	–
$-C(C_2H_5)=CH_2$	7.98	2.47	2.527	–
$-CH(CH_3)-CH=CH_2$	7.71	2.38	2.265	–
$-CH_2-CH_2-CH=CH_2$	7.60	2.35	2.435	–
$-CH=C=CH_2$	8.35	2.60	2.692	–
$-C\equiv CH$	9.64	3.03	3.613	–
$-C\equiv C-CH_3$	9.32	2.92	3.266	–
$-CH_2-C\equiv CH$	8.21	2.55	2.576	–
$-CH_2-C\equiv C-CH_3$	8.14	2.53	2.538	–
$-OH$	11.26	3.58	5.346	5.6
$-CH_2-OH$	8.65	2.70	2.860	–
$-CH_2-CH_2-OH$	8.07	2.50	2.629	–
$-O-CH_3$	10.38	3.28	4.244	–
$-O-CH_2-CH_3$	10.32	3.26	4.124	–
$-NH_2$	8.76	2.74	3.402	5.3
$-NH-CH_3$	8.53	2.66	2.857	–
$-NH(C_2H_5)$	8.51	2.65	2.786	–
$-NH(C_3H_7)$	8.51	2.65	2.775	–
$-NH(C_4H_9)$	8.52	2.65	2.773	–
$-NH(i\text{-}C_3H_7)$	8.50	2.65	2.732	–
$-NH(t\text{-}C_4H_9)$	8.49	2.64	2.690	–
$-N(CH_3)(C_2H_5)$	8.37	2.61	2.469	–
$-N(C_2H_5)_2$	8.36	2.60	2.424	–
$-N(C_3H_7)_2$	8.35	2.60	2.358	–
$-N(i\text{-}C_3H_7)_2$	8.37	2.60	2.411	–
$-N(t\text{-}C_4H_9)_2$	8.34	2.59	2.308	–
$-CHO$	10.08	3.18	3.598	–
$-COCl$	10.70	3.39	3.530	–
$-C(O)CH_3$	9.58	3.01	3.103	–
$-C(O)NH_2$	10.26	3.24	3.219	–
$-COOH$	12.13	3.87	4.504	–
$-O-C(O)-CH_3$	11.69	3.72	4.316	–
$-O-C(O)-CH_2-CH_3$	11.64	3.70	4.290	–
$-C(O)-O-CH_3$	10.91	3.46	3.343	–
$-C(O)-O-C_2H_5$	10.87	3.44	3.324	–
$-SO_3H$	15.67	5.06	2.715	–
$-O-SO_3H$	15.52	5.01	4.157	–
$-SO_3-CH_3$	15.02	4.84	2.594	–

Table 3 (continued)

Group	χ_M	χ_P	$\eta = c_{eff}$	η_{Lit} [b]
$-O-SO_3-CH_3$	15.10	4.87	4.106	–
$-SO_3-CH_2-CH_3$	14.94	4.83	2.578	–
$-O-SO_3-CH_2-CH_3$	15.05	4.85	4.099	–
$-SO_2Cl$	15.20	4.90	2.747	–
$-O-P(O)(O-CH_3)_2$	13.26	4.25	4.223	–
$-P(O)(O-CH_3)_2$	12.14	3.87	2.872	–
$-CN$	11.86	3.78	4.512	5.3
$\underline{C}N^-$	-1.16	-0.60	5.928	
$-\underline{N}C$	11.59	3.69	4.575	–
$-SCN$ (te)[c]	11.13	3.53	2.101	–
(p)[c]	9.66	3.04	2.554	–
$-NCS$ (tr)[c]	11.74	3.74	3.570	–
(p)[c]	11.10	3.52	3.522	–
$-NCO$	12.75	4.08	3.793	–
$-OCN$	13.34	4.27	4.814	–
$-SH$	7.30	2.25	2.963	4.1
$-SeH$	7.08	2.17	2.857	–
$-TeH$	6.83	2.09	3.078	–
$-PH_2$	6.67	2.03	2.375	–
$-AsH_2$	6.47	1.97	1.318	–
$-SbH_2$	6.07	1.83	2.073	–
$-OCl$	12.20	3.89	5.165	4.5
$\underline{Cl}O^-$	2.99	0.80	6.495	
$H_2\underline{O}$	1.77	0.39	4.994	7.0
$\underline{N}H_3$	0.82	0.07	3.994	6.9
$H_2\underline{S}$ (p)[c]	4.94	1.45	2.681	5.3
(te)[c]	4.48	1.30	2.994	
$\underline{P}H_3$ (p)[c]	13.24	4.24	2.911	5.0
(te)[c]	5.03	1.48	2.693	
$\underline{C}O$	7.87	2.44	3.846	6.0
$C\underline{O}$	5.71	1.71	6.722	

[a] Dimensions for χ_M and b^0 are [V], for c, b^1 and η [V/e] and for χ_P [$\sqrt{eV}$]
[b] Experimental values for hardness; taken from Ref. 11
[c] The hybridisation of the sulfur- or phosphorous-atom is given in parentheses
The number in parentheses indicates the skeletal carbon-atom whose orbital electronegativity is listed.
The values for the molecules with underlined atoms are those of the lone pairs in that molecules, calculated with the atomic properties:

Atom	Valence state	b^0	b^1	c	χ_M	χ_P
C	$\underline{tr^2}tr\ \pi$	35.077	18.576	7.718	4.21	1.21
N	$\underline{di^2}di\ \pi\ \pi$	46.909	14.890	9.885	7.37	2.27
	$\underline{tr^2}tr\ tr\ \pi$	44.414	15.409	9.769	5.34	1.59
	$\underline{te^2}te\ te\ te$	42.893	15.507	9.580	4.57	1.33
O	$\underline{tr^2}tr^2tr\ \pi$	53.724	16.520	11.190	8.96	2.81
	$\underline{te^2}te^2te\ te$	51.183	17.630	10.825	7.88	2.44
P	$\underline{s^2}p\ p\ p$	37.452	10.423	6.280	12.33	3.94
	$\underline{te^2}te\ te\ te$	30.014	9.511	5.973	6.12	1.85
S	$s^2\underline{p^2}p\ p$	28.939	11.047	5.962	5.09	1.50
	$\underline{te^2}te^2te\ te$	34.773	12.463	7.045	6.59	2.01
Cl	$\underline{s^2}p^2p^2p$	32.863	12.151	6.495	6.88	2.11

F. Correlations with Molecular Properties

The usefulness of the concepts outlined above depends critically on how well the results obtained will correlate with certain molecular properties. If these correlations are satisfactory, the results, which are readily derived and based solely on atomic data in conjunction with the concept of electronegativity equalization, can be used to predict such molecular properties. Thus far we have tested only a few simple correlations and have found quite satisfactory results.

I. Bond Length

Following a suggestion by Derflinger and Polanski[18] the bond length is approximately given by

$$r_{AB} = r_A + r_B - k|\chi_A - \chi_B| \qquad (29)$$

i.e. as a sum of the covalent radii and a bond shortening proportional to the absolute value of the electronegativity difference. For the 122 bonds investigated by Derflinger and Polanski we have used the calculated group-orbital electronegativities and, via a least

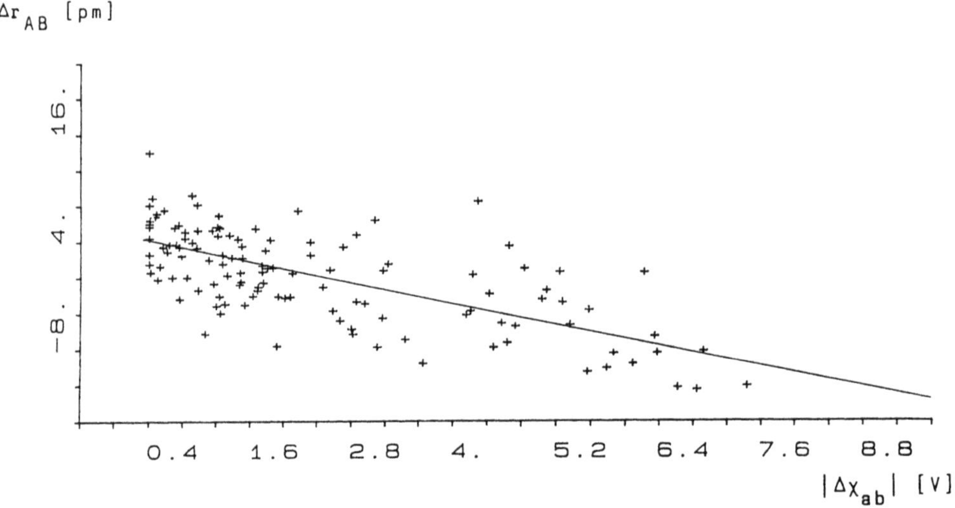

Fig. 1. Plot of Δr_{AB} versus $|\Delta\chi_{AB}|$, where Δr_{AB} is the difference between the internuclear distance r_{AB} and the covalent radii r_A and r_B, while $|\Delta\chi_{AB}|$ is the difference in electronegativities of the groups forming the bond between atoms A and B. The correlation was performed with the equation of Schomaker and Stephenson[26] [see Eq. (29)]:

$$d_{AB} = r_A + r_B - k|\Delta\chi_{AB}|$$

giving the covalent radii (Table 5) and the constant $k = 1.9$ pm/V with a multiple correlation coefficient of 0.998 and a standard deviation of 4.2 pm. 122 bonds from Ref. 18, shown in Table 6 were used to calculate the parameters

Table 4. Bond length's r_{ab} versus electronegativity difference $\Delta\chi_{ab}$[a]

Compound	Bond	r_{ab}[b]	r_{ab}[c]	difference	$\Delta\chi_{ab}$
$SOCl_2$	S–Cl	207.0	194.9	12.1	3.893
F_2	F–F	141.8	131.7	10.1	0.0
SF_6	S–F	158.0	166.8	–8.8	0.669
CF_4	C–F	132.3	140.9	–8.6	1.523
SO_2Cl_2	S–Cl	199.0	191.2	7.8	5.830
AlH	Al–H	164.8	157.0	7.8	4.255
$SiHF_3$	Si–H	145.5	137.9	7.6	2.672
AsF_3	As–F	171.2	178.4	–7.2	3.255
OF_2	O–F	141.8	134.9	6.9	1.766
CHF_3	C–F	132.2	138.6	–6.4	2.707
SiH_3Cl	Si–H	148.3	142.0	6.3	0.505
HI	H–I	161.7	168.0	–6.3	0.849
RbH	Rb–H	236.7	230.7	6.0	4.846
$SiBr_4$	Si–Br	215.0	220.6	–5.6	0.362
BBr	B–Br	188.7	183.1	5.6	4.437
NH_2OH	N–O	146.0	140.4	5.6	2.459
$SiCl_4$	Si–Cl	201.0	206.5	–5.5	0.798
$(PNCl_2)_3$	P–Cl	197.0	202.5	–5.5	2.422
NH_2Cl	N–Cl	177.0	171.7	5.3	0.570
$(SiH_3)_3N$	Si–N	173.8	178.9	–5.1	0.902
SiH_4	Si–H	148.0	142.9	5.1	0.033
SOF_2	S–F	158.5	163.5	–5.0	2.405
CH_3OH	O–H	96.7	101.6	–4.9	3.042
HBr	H–Br	141.4	146.2	–4.8	1.135
SiH_3F	Si–H	146.0	141.4	4.6	0.825
KBr	K–Br	294.0	298.5	–4.5	5.164
C_2H_4	C–H	107.0	111.4	–4.4	0.834
$F_5SO–OSF_5$	S–O	166.0	170.3	–4.3	2.273
KCl	K–Cl	279.0	283.3	–4.3	6.220
$SiHBr_3$	Si–Br	216.0	220.2	–4.2	0.584
H_2	H–H	74.2	70.0	4.2	0.0
AlCl	Al–Cl	213.8	217.9	–4.1	6.446
CH_3SH	C–S	181.8	177.8	4.0	1.262
H_2Se	Se–H	147.0	150.9	–3.9	0.099
$(CH_3)_2Se$	C–Se	197.7	193.8	3.9	0.175
CF_3Br	C–Br	190.8	186.9	3.9	2.308
H_2O	O–H	95.8	99.6	–3.8	4.080
BrF	Br–F	175.6	171.8	3.8	3.831
CH_3Cl	C–Cl	178.1	174.4	3.7	1.917
AlBr	Al–Br	229.5	233.2	–3.7	5.390
CH_3F	C–F	138.5	134.8	3.7	4.692
$SiHCl_3$	Si–Cl	202.1	205.7	–3.6	1.233
HCl	H–Cl	127.5	130.9	–3.4	2.191
CH_4	C–H	109.1	112.5	–3.4	0.274
$NHCl_2$	N–Cl	176.0	172.6	3.4	0.084
CH_2=CHI	C–I	209.2	212.5	–3.3	0.015
$(CH_3)_2S$	C–S	182.0	178.7	3.3	0.816
AsI_3	As–I	255.0	251.8	3.2	0.840
BBr_3	B–Br	187.0	190.1	–3.1	0.767
NH_3	N–H	101.5	104.6	–3.1	1.621
$(CH_3)_2O$	C–O	142.0	145.1	–3.1	2.768
SiF_3Br	Si–Br	215.3	218.4	–3.1	1.537
NH	N–H	103.8	106.9	–3.1	0.448
$SbBr_3$	Sb–Br	251.0	248.0	3.0	1.442

Table 4 (continued)

Compound	Bond	r_{ab}[b]	r_{ab}[c]	difference	$\Delta\chi_{ab}$
Cl_2O	O–Cl	170.1	167.1	3.0	2.834
CBr_4	C–Br	194.2	191.2	3.0	0.067
SO_2F_2	S–F	157.0	160.0	–3.0	4.241
BCl_3	B–Cl	173.0	175.9	–2.9	1.287
SiH_2Cl_2	Si–Cl	202.0	204.8	–2.8	1.686
PI_3	P–I	243.0	240.2	2.8	0.750
KI	K–I	323.0	320.3	2.7	4.878
$AsBr_3$	As–Br	233.0	230.3	2.7	0.954
$(CH_3)_2NH$	C–N	146.0	148.6	–2.6	1.075
$(CH_3)_3P$	C–P	187.0	184.4	2.6	0.350
RbI	Rb–I	326.0	328.6	–2.6	5.695
RbCl	Rb–Cl	289.0	291.6	–2.6	7.037
C_2H_5F	C–F	137.5	134.9	2.6	4.640
CI_4	C–I	215.0	212.5	2.5	0.007
SbI_3	Sb–I	267.0	269.5	–2.5	1.298
H_2S	S–H	134.6	137.1	–2.5	0.125
CH_3I	C–I	213.9	211.4	2.5	0.575
I_2	J–I	266.7	269.1	–2.4	0.0
BrCl	Br–Cl	213.8	211.4	2.4	1.056
NaI	Na–I	290.0	287.6	2.4	5.184
SiH_3I	Si–I	243.3	241.0	2.3	0.816
CH_3SH	S–H	132.9	135.2	–2.3	1.090
$CHBr_3$	C–Br	193.0	190.7	2.3	0.302
NF_3	N–F	137.1	134.9	2.2	1.921
ClF	Cl–F	162.8	160.6	2.2	2.775
K_2	K–K	392.3	390.2	2.1	0.0
KH	K–H	224.4	222.4	2.0	4.029
$(CH_3)_3As$	C–As	198.0	196.0	2.0	0.424
CH_3COF	C–F	137.0	138.9	–1.9	2.561
$(CH_3)_3N$	C–N	147.0	148.9	–1.9	0.928
C_2H_2	C–H	106.4	108.3	–1.9	2.463
$CH_2{=}CHCl$	C–Cl	173.6	175.5	–1.9	1.357
NaBr	Na–Br	264.0	265.8	–1.8	5.470
CH_2Cl_2	C–Cl	177.2	175.4	1.8	1.382
Br_2	Br–Br	228.4	226.7	1.7	0.0
SiH_3Br	Si–Br	220.9	219.2	1.7	1.102
Cl_2	Cl–Cl	198.8	200.2	–1.4	0.0
SCl_2	S–Cl	199.0	200.3	–1.3	1.085
SiI_4	Si–I	243.0	241.7	1.3	0.424
SiH_3Cl	Si–Cl	205.0	203.9	1.1	2.158
$SbCl_3$	Sb–Cl	232.5	233.5	–1.0	2.074
NaH	Na–H	188.7	189.7	–1.0	4.335
PH_3	P–H	142.1	141.1	1.0	0.509
H_2O_2	O–H	97.0	96.0	1.0	5.951
LiH	Li–H	159.5	160.4	–0.9	4.171
CH_2F_2	C–F	135.8	136.6	–0.8	3.756
RbBr	Rb–Br	306.0	306.8	–0.8	5.981
CCl_4	C–Cl	176.6	177.3	–0.7	0.383
PBr_3	P–Br	218.0	218.7	–0.7	0.873
CH_3NH_2	C–N	147.4	148.1	–0.7	1.347
CH_3COCl	C–Cl	177.0	177.7	–0.7	0.214
AsH_3	As–H	151.9	152.5	–0.6	0.709
Li_2	Li–Li	267.3	266.8	0.5	0.0
CH_2Br_2	C–Br	190.7	190.2	0.5	0.565

Table 4 (continued)

Compound	Bond	r_{ab} [b]	r_{ab} [c]	difference	$\Delta\chi_{ab}$
NaCl	Na–Cl	251.0	250.6	0.4	6.526
SbH_3	Sb–H	170.7	170.3	0.4	1.108
$B(CH_3)_3$	B–C	156.0	155.6	0.4	0.320
HF	H–F	91.7	91.3	0.4	4.966
CH_3OH	C–O	142.8	143.1	–0.3	3.806
$CHCl_3$	C–Cl	176.7	176.4	0.3	0.871
CHI_3	C–I	212.0	212.2	–0.2	0.162
CH_3SiH_3	C–Si	185.7	185.5	0.2	0.241
$TeBr_2$	Te–Br	251.0	250.8	0.2	0.983
$TeCl_2$	Te–Cl	236.0	236.2	–0.2	1.708
PCl_3	P–Cl	204.3	204.5	–0.2	1.408
CH_2I_2	C–I	212.0	211.9	0.1	0.355
ICl	I–Cl	232.1	232.1	0.0	1.342
$AsCl_3$	As–Cl	216.1	216.1	0.0	1.476

Multiple Corr. Coeff.: 0.998
Std. Dev.: 4.2 [pm]
k: 1.9 ± 0.2 [pm/V]

[a] The length's were computed with the equation of Shomaker and Stephenson[26)]:

$$r_{ab} = r_a + r_b - k \cdot |\chi_a - \chi_b| ,$$

where r_a and r_b are the covalent radii for atom A and B. These were evaluated in the same process (see Table 5)

[b] Experimental bond length's were taken from Ref. 18

[c] Bond length's from regression.
Bond length's are given in [pm], electronegativity differences $\Delta\chi$ in [V]

square regression, determined the covalent radii and the proportionality constant k in Eq. (29). The results are given in Table 4 and Fig. 1 with the covalent radii obtained in Table 5. Though the correlation coefficient between $\Delta r_{AB} = (r_{AB} - r_A - r_B)$ and $\Delta\chi_{AB}$ of $r = 0.68$ for this simple relation is not especially good, the relation is still quite satisfactory. As the standard deviation of the absolute error indicates, a prediction of the bond distance with an error of about 4 pm is achieved readily using Eq. (29) with $k = 1.9$ (pm/V). A more detailed analysis, which is in progress, hopefully will yield an even more satisfactory relation.

II. NMR-Chemical Shifts

The chemical shift δ of an NMR resonance of an atomic nucleus depends on the charge distribution around this nucleus as well as on the bulk para- and diamagnetic influences of the molecule and its surrounding. The latter effects should be roughly the same for two neighbouring atoms. Thus we should expect in molecules of the type $R–CH_2–CH_3$ that the difference in the ^{13}C-chemical shift of the carbon in CH_2 and CH_3 correlates with the electron withdrawing power, electronegativity, of the molecular group R to give the relation (19)

Table 5. Covalent radii[a] in [pm]

Element	r	Error[b]	r_{lit}
H	35.0	0.9	32.0
Li	133.4	1.9	134.0
B	78.3	2.1	82.0
C	78.0	0.8	77.0
N	72.7	1.4	74.0
O	72.4	1.5	70.0
F	65.8	1.0	68.0
Na	163.0	2.1	154.0
Al	130.2	2.5	126.0
Si	108.0	1.1	117.0
P	107.1	1.8	110.0
S	102.3	1.3	104.0
Cl	100.1	0.8	99.0
K	195.1	1.5	196.0
As	118.8	1.8	119.0
Se	116.1	3.0	116.0
Br	113.3	0.9	114.0
Rb	205.0	2.1	216.0
Sb	137.4	2.1	138.0
Te	139.4	3.0	135.0
I	134.6	1.0	133.0

[a] Radii were evaluated with the 122 bonds of Table 4. The literature data are those given by Sanderson[9)]

[b] Std. Error of regression

$$\chi_R = a + b[\delta(CH_3) - \delta(CH_2)] = a + b \cdot \Delta\delta \tag{30}$$

The results of correlating the calculated group electronegativities with $\Delta\delta$-chemical shift values for the ^{13}C-resonances are presented in Table 6 and Fig. 2. The correlation is indeed satisfactory.

III. Proton Affinities

The proton affinity defined as the negative of the reaction enthalpy of the gas-phase reaction

$$M + H^+ \rightleftarrows MH^+ \tag{31}$$

and not easily determined directly. Relative values may be obtained out of the determination of the equilibrium constants for proton-exchange reactions of the type

$$RH^+ + M \rightleftarrows MH^+ + R \tag{32}$$

Recently standardized "absolute" values have been compiled[20)]. We have correlated these values of nitrogen-containing compounds with (a) the charge on nitrogen and (b)

Table 6. Comparison of the ^{13}C-chemical shifts with electronegativity of several ethyl compounds

Structure	χ^a	χ^b	difference	$\Delta\delta\ ^{13}C^c$
H H O H-C-C-S-O-H H H O	15.519	11.934	3.585	38.600
H H O H H-C-C-S-O-C-H H H O H	15.022	11.639	3.383	36.000
H H O H H H-C-C-P-O-C-C-H H H O H H H-C-H H-C-H H	12.066	8.977	3.089	12.500
H H S S H H H-C-C-O-C-S-S-C-O-C-C-H H H H H	11.330	14.109	-2.779	57.800
H H O O H H H-C-C-S-P-S-S-S-P-S-C-C-H H H S S H H H-C-H H-C-H H-C-H H-C-H H H	10.419	13.101	-2.682	48.900
H H O H H H-C-C-O-S-O-C-C-H H H O H H	15.047	12.410	2.637	42.800
H H H-C-C-I H H	8.026	10.042	-2.016	21.900
H H O H H-C-C-S-N H H O H	14.172	12.172	2.000	40.700
H H H O H H H H-C-C-C-S-C-C-C-H H H H O H H H	9.726	7.867	1.859	2.700
H H O H-C-C-S-Cl H H O	15.198	13.350	1.848	51.100

Table 6 (continued)

Structure	χ^a	χ^b	difference	$\Delta\delta$ $^{13}C^c$
H H H H H H H H H-C-C-C-C-H H-C-C-C-P H H H H H H H H-C-H H-C-H H-C-H H-C-H H	7.586	9.158	−1.572	14.100
H H O H H H-C-C-O-C-S-C-C-O-H H H H H	11.663	13.158	−1.495	49.400
H H O H H-C-C-S-C-C-H H H H	9.766	8.558	1.208	8.800
H H H-C-C-S-H H H	8.713	7.629	1.084	0.600
H H O H H H H-C-C-O-P-N-C-C-H H H O H H H-C-H H-C-H H	10.702	9.713	0.989	19.000
H H O H H H H H H H-C-C-O-P-N-C-C-C-C-C-C-H H H O H H H H H H H-C-H H-C-H H	7.599	8.546	−0.947	8.700
H H H H H H-C-C-C-C-C-Cl H H H H H	7.597	8.512	−0.915	8.400
H H H H H H H H H H H-C-C-C-C-C-S-S-C-C-C-C-C-H H H H H H H H H H H	7.628	8.524	−0.896	8.500
H H H H H H-C-C-C-C-C-S-C•N H H H H H	7.655	8.501	−0.846	8.300
H H H-C-C-O-H H H	11.257	12.036	−0.779	39.500

Table 6 (continued)

Structure	χ^a	χ^b	difference	$\Delta\delta$ $^{13}C^c$
H H H H H-C-C-C-C-S-H H H H H	7.705	8.478	−0.773	8.100
H H S H H H-C-C-O-P-O-C-C-H H H Cl H H	12.523	13.271	−0.748	50.400
H H H H H H H H H-C-C-C-S-C-C-S-C-C-C-H H H H H H H H H	7.929	8.660	−0.731	9.700
H H H H H O H H H H H H-C-C-C-C-C-O-S-O-C-C-C-C-C-H H H H H H O H H H H H	7.812	8.512	−0.700	8.400
H H H H S H H H H H-C-C-C-C-S-C-S-C-C-C-C-H H H H H H H H H	7.824	8.524	−0.700	8.500
H H H H H H H H H-C-C-C-C-S-S-C-C-C-C-H H H H H H H H H	7.775	8.467	−0.692	8.000
H H H H H-C-C-C-C-S-C≡N H H H H	7.847	8.444	−0.597	7.800
H H H-C-C-S-C≡N H H	9.659	9.067	0.592	13.300
H H O H H S H H-C-C-O-C-N-N-C-N H H H	12.241	12.818	−0.577	46.400
H H H H H-C-C-S-C-C-H H H H H	8.231	8.773	−0.542	10.700
H H O H H H-C-C-O-P-O-C-C-H H H O H H H	13.451	12.942	0.509	47.500

Table 6 (continued)

Structure	χ^a	χ^b	difference	$\Delta\delta$ $^{13}C^c$
H H H H H H H-C-C-S-C-C-S-C-C-H H H H H H H	8.333	8.830	–0.497	11.200
H H O H H-C-C-O-S-C-H H H O H	13.937	13.452	0.485	52.000
H H O Cl H-C-C-O-P-C-Cl H H O Cl H-C-H H-C-H H	12.839	13.294	–0.455	50.600
H H S H H H-C-C-O-P-O-C-C-H H H O H H H-C-H H-C-H H	12.563	13.010	–0.447	48.100
H H-C-H H-C-H H H S H H H-C-C-S-C-S-C-C-H H H H H H	8.537	8.864	–0.327	11.500
H H H H O H H H H H-C-C-C-C-S-C-C-C-C-H H H H H H H H H	8.117	7.867	0.250	2.700
H H Cl H-C-C-Si-Cl H H Cl	8.570	8.807	–0.237	11.000
H H H H O H H H H H-C-C-C-C-O-S-O-C-C-C-C-H H H H H H H H H	8.384	8.184	0.200	5.500
H H O H H H H-C-C-O-P-C-C=C H H O H H H-C-H H-C-H H	12.528	12.727	–0.199	45.600

Table 6 (continued)

Structure	χ[a]	χ[b]	difference	Δδ [13]C[c]
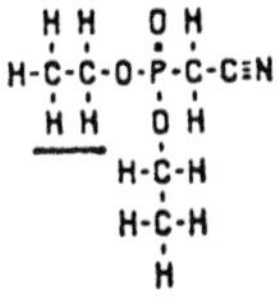 H H O H H-C-C-O-P-C-C≡N H H O H H-C-H H-C-H H	12.791	12.954	–0.163	47.600
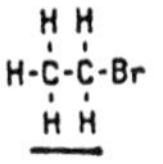 H H H-C-C-Br H H	8.312	8.467	–0.155	8.000
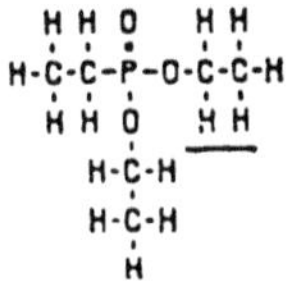 H H O H H H-C-C-P-O-C-C-H H H O H H H-C-H H-C-H H	12.483	12.625	–0.142	44.700
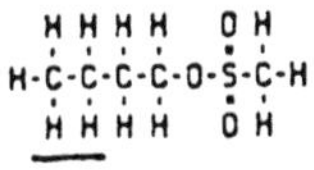 H H H H O H H-C-C-C-C-O-S-C-H H H H H O H	8.268	8.161	0.107	5.300
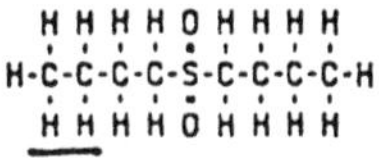 H H H H O H H H H H-C-C-C-C-S-C-C-C-C-H H H H H O H H H H	8.389	8.490	–0.101	8.200
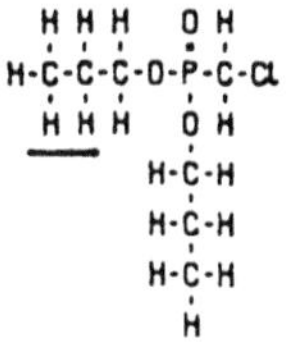 H H H O H H-C-C-C-O-P-C-Cl H H H O H H-C-H H-C-H H-C-H H	9.057	9.147	–0.090	14.000
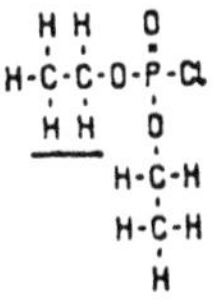 H H O H-C-C-O-P-Cl H H O H-C-H H-C-H H	13.201	13.248	–0.047	50.200
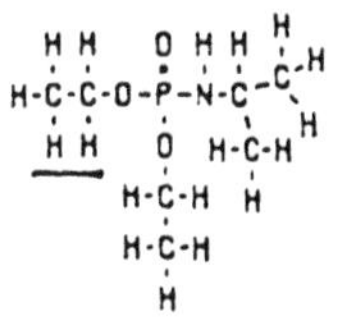 H H O H H H H-C-C-O-P-N-C-C-H H H O H-C-H H H-C-H H H-C-H H	12.746	12.704	0.042	45.400
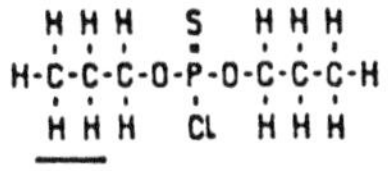 H H H S H H H H-C-C-C-O-P-O-C-C-C-H H H H Cl H H H	9.017	9.056	–0.039	13.200

Table 6 (continued)

Structure	χ^a	χ^b	difference	$\Delta\delta$ $^{13}C^c$
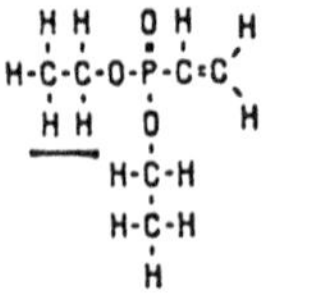 H H O H H H-C-C-O-P-C=C H H O H H-C-H H-C-H H	12.654	12.693	–0.039	45.300
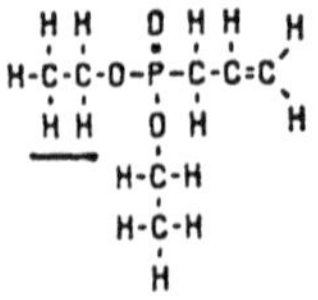 H H O H H H H-C-C-O-P-C-C=C H H O H H H-C-H H-C-H H	12.758	12.727	0.031	45.600
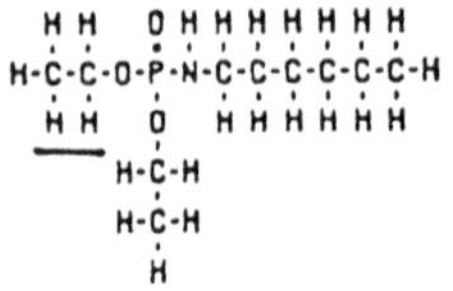 H H O H H H H H H H-C-C-O-P-N-C-C-C-C-C-C-H H H O H H H H H H H-C-H H-C-H H	12.757	12.727	0.030	45.600
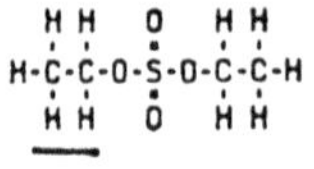 H H O H H H-C-C-O-S-O-C-C-H H H O H H	13.799	13.803	–0.004	55.100
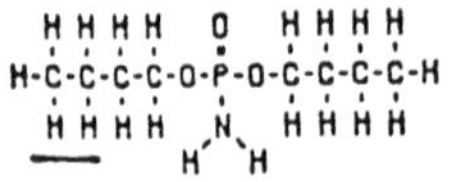 H H H H O H H H H H-C-C-C-C-O-P-O-C-C-C-C-H H H H H N H H H H H H	8.140	8.138	0.002	5.100

Corr. Coeff.: 0.861
Std. Dev.: 1.307 [V]
Coefficient: 7.561 ± 0.300 [V]
Slope: 0.113 ± 0.009 [V/ppm]

a Theoretical electronegativity in [V]
b Electronegativity given by regression in [V]
c The underlined ethyl group values in [ppm] are taken from[27]

with the orbital electronegativity of the lone pair on nitrogen in these compounds. The correlation (b)

$$PA = a + b\chi_N \tag{33}$$

is much superior to the correlation (a)[16]. This indicates that not only is the charge on the nitrogen important for the proton affinity but also the hybridization of the orbital on nitrogen, which forms the bond to the proton. The results of this correlation are presented in Table 7 and Fig. 3 indicating that the proton affinities of ammines may be predicted using relation (33) with a reliability of about ± 27 [kJ/mol], a quite gratifying result.

Table 7. Proton affinities PA versus electronegativity χ_N of lone pairs in several nitrogen compounds[a]

Compound	PA[b]	PA[c]	difference	χ_N
$NC{-}CH_2\underline{NH}_2$	826.000	890.293	–64.293	1.281
$\underline{NH}_3$	853.500	917.638	–64.138	0.819
$(CF_3){-}CH_2\underline{NH}_2$	847.000	894.851	–47.851	1.204
$(i\text{-}C_3H_7)_2(C_2H_5)\underline{N}$	984.000	936.518	47.482	0.500
$NC{-}(CH_2)_2{-}\underline{NH}_2$	866.000	912.548	–46.548	0.905
$i\text{-}C_3H_7C\underline{N}$	813.000	766.651	46.349	3.370
$(t\text{-}C_4H_9)_2\underline{NH}$	976.000	932.790	43.210	0.563
$NC{-}CH_2{-}\underline{NH}(CH_3)$	862.000	901.657	–39.657	1.089
$(CHF_2){-}CH_2\underline{NH}_2$	868.000	906.274	–38.274	1.011
$(s\text{-}C_4H_9)_2\underline{NH}$	970.000	932.198	37.802	0.573
$(C_2H_5)_3\underline{N}$	972.000	936.459	35.541	0.501
$(C_2H_5)_2(n\text{-}C_3H_7)\underline{N}$	971.000	936.223	34.777	0.505
$(CH_3)_2(t\text{-}C_4H_9)\underline{N}$	971.000	936.400	34.600	0.502
$H_2N{-}\underline{NH}_2$	856.000	890.412	–34.412	1.279
$CH_2{=}C(CH_3)\underline{NH}_2$	947.000	912.784	34.216	0.901
$(CF_3){-}(CH_2)_2{-}\underline{NH}_2$	881.000	914.086	–33.086	0.879
$(CH_3)_3Si{-}CH_2{-}\underline{N}{-}(CH_3)_2$	968.000	937.525	30.475	0.483
$(i\text{-}C_3H_7)_2\underline{NH}$	963.000	932.612	30.388	0.566
$CH_3\underline{NH}_2$	896.000	926.220	–30.220	0.674
$(CH_2F){-}CH_2\underline{NH}_2$	888.000	916.809	–28.809	0.833
$(CF_3){-}CH_2{-}\underline{NH}(CH_3)$	878.000	905.386	–27.386	1.026
$(CH_3)_2(neo\text{-}C_5H_{11})\underline{N}$	962.000	935.808	26.192	0.512
$(CH_3)(C_2H_5)_2\underline{N}$	962.000	936.400	25.600	0.502
$(CH_3)_2(i\text{-}C_3H_7)\underline{N}$	961.000	935.808	25.192	0.512
$(i\text{-}C_4H_9)_2\underline{NH}$	956.000	931.547	24.453	0.584
$(n\text{-}C_4H_9)_2\underline{NH}$	956.000	931.606	24.394	0.583
$CH_3{-}C\underline{N}$	788.000	763.987	24.013	3.415
$HC{\equiv}C{-}CH_2{-}\underline{NH}_2$	882.000	905.919	–23.919	1.017
$(CF_3)C\underline{N}$	695.000	673.253	21.747	4.948
$(n\text{-}C_3H_7)_2\underline{NH}$	952.000	931.902	20.098	0.578
$CH_3C(H){=}\underline{N}(C_2H_5)$	932.000	912.548	19.452	0.905
$(CH_3)_2N{-}\underline{NH}_2$	920.000	901.361	18.639	1.094
$(C_2H_5)\underline{NH}_2$	908.000	926.575	–18.575	0.668
$(C_2H_5)(i\text{-}C_3H_7)\underline{NH}$	951.000	932.494	18.506	0.568
$CH_2{=}C(H){-}CH_2\underline{NH}_2$	903.000	920.656	–17.656	0.768
$(CH_3)_2(C_2H_5)\underline{N}$	952.000	936.341	15.659	0.503
$(CH_3)N{=}\underline{N}(CH_3)$	866.000	881.415	–15.415	1.431
$Cl{-}C\underline{N}$	735.000	720.011	14.989	4.158
$n\text{-}C_3H_7\underline{NH}_2$	912.000	926.279	–14.279	0.673
$Br{-}C\underline{N}$	746.000	760.140	–14.140	3.480
$Cl{-}CH_2{-}C\underline{N}$	751.000	764.106	–13.106	3.413
$(C_2H_5)_2\underline{NH}$	945.000	932.375	12.625	0.570
$n\text{-}C_4H_9\underline{NH}_2$	914.000	926.161	–12.161	0.675
$i\text{-}C_3H_7\underline{NH}_2$	915.000	926.871	–11.871	0.663
$F{-}(CH_2)_3{-}\underline{NH}_2$	911.000	922.550	–11.550	0.736
$i\text{-}C_4H_9\underline{NH}_2$	915.000	926.101	–11.101	0.676
$(CCl_3){-}C\underline{N}$	735.500	746.290	–10.790	3.714
$n\text{-}C_6H_{13}\underline{NH}_2$	916.000	926.042	–10.042	0.677
$n\text{-}C_7H_{15}\underline{NH}_2$	916.000	926.042	–10.042	0.677
$n\text{-}C_5H_{11}\underline{NH}_2$	916.000	926.042	–10.042	0.677
$NC{-}CH_2{-}C\underline{N}$	735.000	726.108	8.892	4.055
$HC\underline{N}$	717.000	725.516	–8.516	4.065
$neo\text{-}C_5H_{11}\underline{NH}_2$	917.500	925.924	–8.424	0.679

Table 7 (continued)

Compound	PA[b]	PA[c]	difference	χ_N
$(CH_3)_3\underline{N}$	942.000	936.341	5.659	0.503
s-$C_4H_9\underline{N}H_2$	922.000	926.634	−4.634	0.667
$(CH_3)_3Si{-}\underline{N}(CH_3)_2$	946.000	941.490	4.510	0.416
n-$C_8H_{17}\underline{N}H_2$	922.000	926.042	−4.042	0.677
t-$C_5H_{11}\underline{N}H_2$	930.000	926.871	3.129	0.663
t-$C_4H_9\underline{N}H_2$	924.000	927.108	−3.108	0.659
c-$C_6H_{11}\underline{N}H_2$	925.500	925.806	−0.306	0.681
$(CH_3)(C_2H_5)\underline{N}H$	932.000	932.198	−0.198	0.573

Corr. Coeff. 0.929
Std. Dev.: 27.483 [kJ/mol]
Coefficient: 966.112 ± 5.077 [kJ/mol]
Slope: −59.187 ± 3.080 [kJ/mol V]

[a] The basic center is underlined
[b] Experimental proton affinities are from Ref. 20
[c] Data given by regression

Dimensions are [kJ/mol] for PA and [V] for χ_N

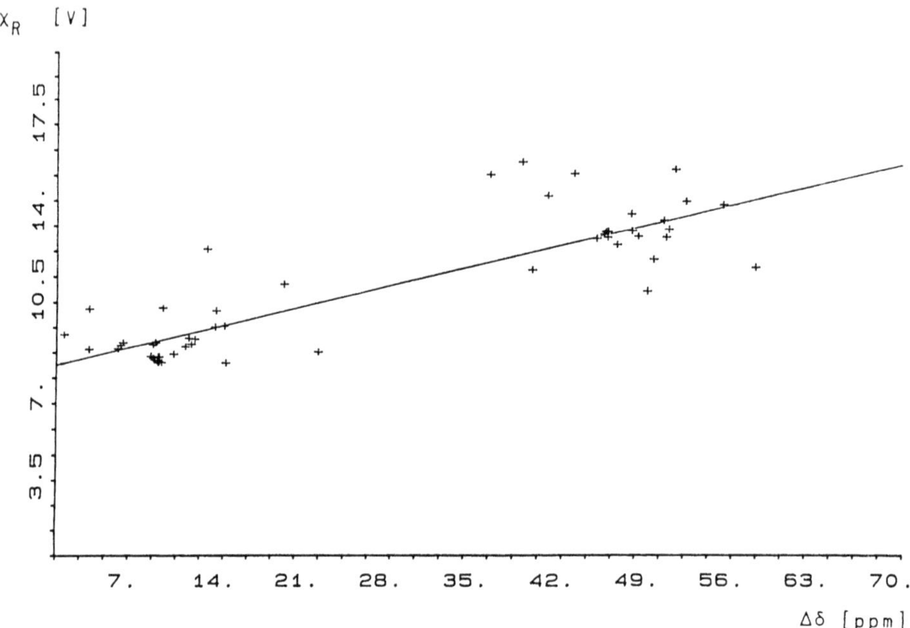

Fig. 2. Plot of the group electronegativity χ_R versus the difference of the ^{13}C-chemical-shifts in NMR-spectra of the CH_2- and CH_3-carbon atoms in several ethyl compounds:

$CH_3{-}CH_2{-}R$

The intercept and slope, a and b, for Eq. (30)

$$\chi_R = a + b\,\Delta\delta$$

were determined by regression to be 7.561 V and 0.113 V/ppm respectively. The correlation was made with the values of Table 6, taken from Ref. 27, which yield a correlation coefficient of 0.861 and a standard deviation of 1.307 V

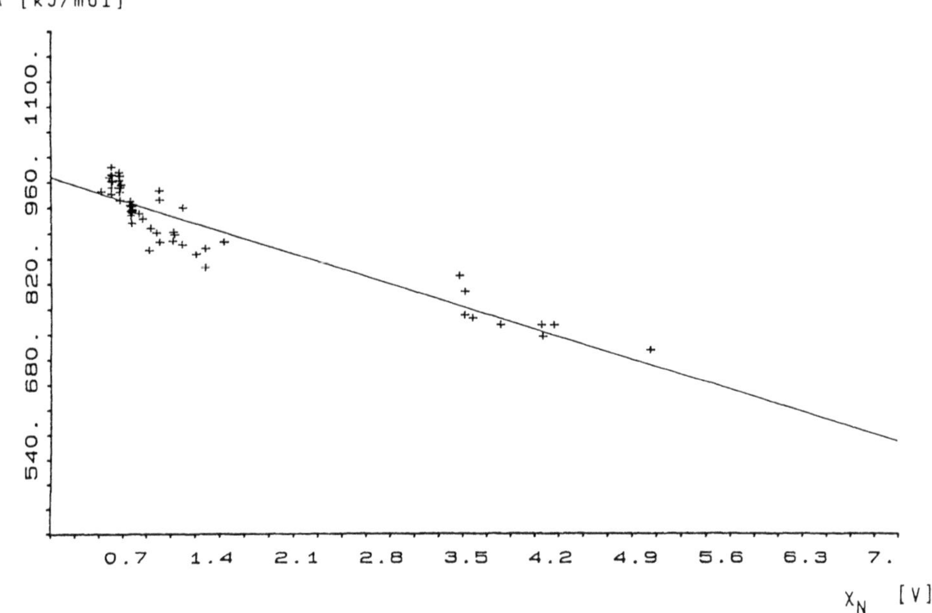

Fig. 3. The experimental proton affinity PA (20) versus electronegativity of lone pairs χ_N of the basic center in several nitrogen compounds is shown. The fit from the correlation of these values by Eq. (33):

$$PA = a + b\chi_N$$

with a = 966.112 kJ/mol and b = −59.187 kJ/mol V and a correlation coefficient of 0.929 in conjunction with a standard deviation of 27 kJ/mol is very good

Certainly the correlations between orbital electronegativities and molecular properties presented here can still be extended and refined. We will attempt to do this in the future. We will conclude here with an investigation we have carried out to see whether the increment method[21)] for the determination of group electronegativities can be justified. To this end, we have calculated the electronegativity on the α, β, γ, ... positions on α-halogen substituted n-octane. The results obtained are given in Fig. 4. A fit of an exponential function to the data indicates that the inductive effect of the electron-withdrawing groups is reduced by a factor of 1/3 by each intervening bond, justifying the increment method.

Appendix A

We will present here first a rigorous quantum-mechanical derivation of the orbital electronegativity

$$\chi_i = \frac{\partial E}{\partial q_i} = -\frac{\partial E}{\partial n_i} \tag{A1}$$

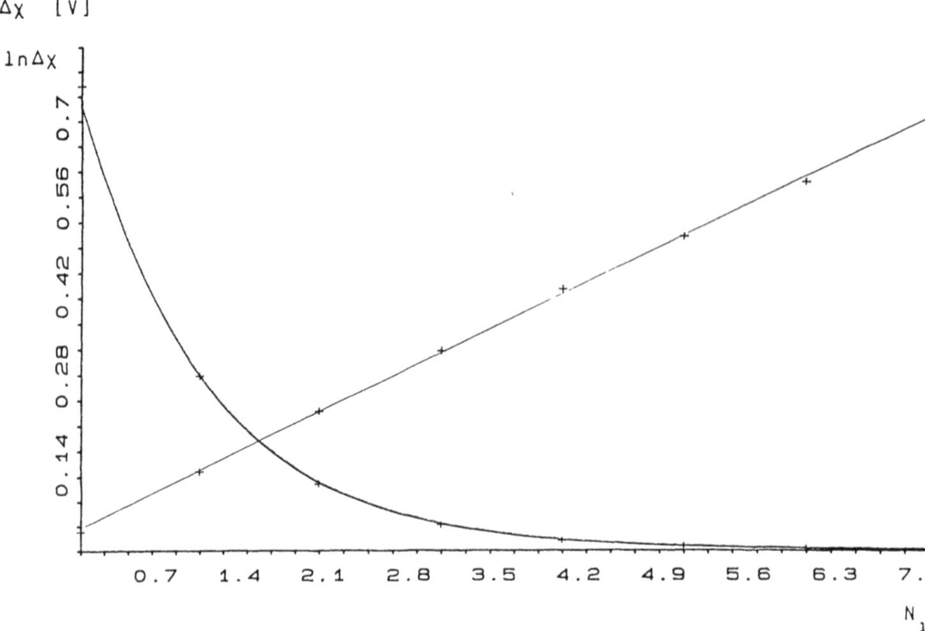

Fig. 4. Decrease in electronegativity $\Delta\chi$ as a function of successive inserting of a CH_2-group, where χ is the difference in electronegativity of octylfluoride compared with octane at the same position of the carbon skeleton

(H) F–C–C–C–C–C–C–C–C– .

N_1 gives the number of inserted CH_2-groups (1–7) between the carbon atom whose group-orbital electronegativity was calculated and the one (1), which forms the bond to F or H. Also shown in the figure is the negative of the natural logarithm of those $\Delta\chi$ values, divided by 10. The plotted points are the calculated values. The correlation with the equation

$\Delta\chi = a\,e^{-bN_1}$

yield the values a = 0.707 V and b = 0.929 with a correlation coefficient of 1.000

This will be followed by a critical examination of the meaning of the "chemical potential" derived, using density functional theory, and identified by Parr et al.[5] with electronegativity and orbital electronegativity. We will show that this identification is erroneous, if the term electronegativity is to retain its commonly accepted meaning. For these discussions we will require some mathematical detail.

To obtain an explicit expression of the partial derivative of the energy of the state Ψ (valence state) of a system (atom) with respect to an orbital occupation number n_i it is convenient to work in the general, particle-independent, Fock space and make use of the creation and annihilation operators a_i^+ and a_i respectively, which create or annihilate a particle characterized by the orbital i. The orbital occupation number in state Ψ is defined as

$$n_{0,i} = \langle\Psi|a_i^+a_i|\Psi\rangle \tag{A2}$$

and the energy of the state Ψ is the extremum of the expectation value

$$E = \langle E \rangle = \langle \Psi | H | \Psi \rangle \tag{A3}$$

$$\langle \delta\Psi | H | \Psi \rangle = 0 \tag{A4}$$

provided the state function is and remains normalized, i.e.

$$\langle \Psi | \Psi \rangle = 1 \tag{A5}$$

Note that the "valence-state" energy, which is not the energy of a conventional stationary state can be characterized in this way by using additional constraints appropriate for the valence state. These constraints are analogous to those necessary if excited state energies are desired.

In order to move infinitesimally from the N-particle subspace of Fock space to the (N − 1)-particle subspace we introduce a parameter λ and the function $(1 + \lambda a)\Psi$. Here and in the following we drop the index i to keep the notation simple. It is understood that we are dealing with a particular orbital i, which may be any orbital of the system, natural or not. We write

$$\frac{\partial \langle E \rangle}{\partial n} = \frac{\partial \langle E \rangle}{\partial \lambda^2} \frac{\partial \lambda^2}{\partial n} \tag{A6}$$

and examine n as a function of λ, which gives

$$n(\lambda) = \frac{\langle (1 + \lambda a)\Psi | a^+ a | (1 + \lambda a)\Psi \rangle}{\langle (1 + a\lambda)\Psi | (1 + a\lambda)\Psi \rangle} = \frac{n_0}{1 + \lambda^2 n_0} \tag{A7}$$

From this we obtain

$$\lambda^2 = \frac{1}{n(\lambda)} - \frac{1}{n_0}$$

and

$$\frac{\partial \lambda^2}{\partial n} = \frac{-1}{n^2(\lambda)} = -\left(\frac{1 + \lambda^2 n_0}{n_0}\right)^2 \tag{A8}$$

For the other partial derivative on the right hand side of Eq. (A6) we get

$$\begin{aligned} \frac{\partial \langle E \rangle}{\partial \lambda^2} &= \frac{\partial}{\partial \lambda^2} \frac{\langle (1 + \lambda a)\Psi | H | (1 + \lambda a)\Psi \rangle}{\langle (1 + \lambda a)\Psi | (1 + \lambda a)\Psi \rangle} \\ &= \frac{\langle \Psi | a^+ H a | \Psi \rangle - n_0 \langle \Psi | H | \Psi \rangle}{(1 + \lambda^2 n_0)^2} \end{aligned} \tag{A9}$$

Substituting (A9) in Eq. (A6) yields

$$\left.\frac{\partial\langle E\rangle}{\partial n}\right|_{n_0} = \frac{1}{n_0}\langle\Psi|H|\Psi\rangle - \frac{1}{n_0^2}\langle\Psi|a^+Ha|\Psi\rangle \tag{A10}$$

the desired result, which can be identified with the orbital electronegativity as we have done in Eq. (3). The right-hand side of Eq. (A10) may be explicated further by writing

$$\langle\Psi|a^+Ha|\Psi\rangle = \langle a^+a\Psi|H|\Psi\rangle + \langle\Psi|a^+[H,a]|\Psi\rangle \tag{A11}$$

which yields

$$\left.\frac{\partial\langle E\rangle}{\partial n}\right|_{n_0} = \frac{1}{n_0^2}[\langle(n_0 - a^+a)\Psi|H|\Psi\rangle + \langle\Psi|a^+[a,H]|\Psi\rangle] \tag{A12}$$

Here we have changed the order and sign of the commutator. We expect the first term $(n_0 - a^+a)\Psi|H|\Psi\rangle$ to be small. It will be zero if: (i) Ψ is an eigenfunction of H or (ii) Ψ is an eigenfunction of a^+a, i.e. a single Slater determinant.

To progress further we need to specify the Hamiltonian.

$$H = \sum_{pq} h_{pq}a_p^+a_q + 1/2\sum_{pqrs} g_{pq,rs}a_p^+a_r^+a_sa_q \tag{A13}$$

with the one and two particle integrals over the orbitals as

$$\begin{aligned} h_{pq} &= \langle p|h|q\rangle \\ g_{pq,rs} &= \left\langle p\left\langle r\right|\frac{1}{r_{12}}\left|s\right\rangle q\right\rangle \end{aligned} \tag{A14}$$

If we examine the operator $a_i^+[a_i, H]$, using the anticommutation relations of the creation and annihilation operators, we obtain

$$a_i^+[a_i,H] = \sum_q\left[h_{iq}a_i^+a_q + \sum_{rs} g_{iq,rs}a_i^+a_r^+a_sa_q\right] \tag{A15}$$

and therefore

$$\langle\Psi|a_i^+[a_i,H]|\Psi\rangle = \sum_q\left[h_{iq}\gamma_{iq} + \sum_{rs} g_{iq,rs}\Gamma_{iq,rs}\right] \tag{A16}$$

where we used the reduced density-matrix elements

$$\gamma_{iq} = \langle\Psi|a_i^+a_q|\Psi\rangle$$

and (A17)

$$\Gamma_{iq,rs} = \langle\Psi|a_i^+a_r^+a_sa_q|\Psi\rangle$$

We recognize the right-hand side of Eq. (A16) as the orbital energy ε_{ii}, i.e. the diagonal Lagrange multiplier of the generalized Fock equations obtained in MCSCF theory[23] [see Eqs. (21) and (35) of Ref. 23]. Note that ε_{ii} here is n_0 times larger than the conventional orbital energies, where the Fock operators are normalized to a single occupation and not to γ_{ii}. The derivation thus far has been general for spin-orbitals. An extension to spatial orbitals is straightforward as long as H is spin-independent. For valence states, where we have a spin-averaged spatial orbital occupation between 0 and 2 we just have to use $(1 + \lambda[a_\alpha + a_\beta])\Psi$ and the spin-free density operators

$$\gamma_{pq} = \sum_{\sigma} a^{+}_{p\sigma} a_{q\sigma}$$

and (A18)

$$\Gamma_{pq,rs} = \sum_{\sigma,\varrho} a^{+}_{p\sigma} a^{+}_{r\varrho} a_{s\varrho} a_{q\sigma}$$

in Eqs. (A2), (A7), (A9) through (A13) and (A15) through (A17). We will forego here a more detailed discussion, as it would also require an explicit definition of the valence state of an atom. This will be presented elsewhere[24]. We will rather focus on the discrepancy between our result and that obtained by Parr et al.[5].

We obtained here the orbital electronegativity as

$$\chi_i(n_{0,i}) = -\left.\frac{\partial E}{\partial n_i}\right|_{n_{0,i}} = \left.\frac{\partial E}{\partial q_i}\right|_{q_{0,i}} \tag{A19}$$

to be equal to the orbital energy of the valence-state orbital i with the occupation n_{0i}, which is different for the different orbitals of a system (atom). This is in contradiction to what Parr et al.[5] found for the chemical potential μ, which they had identified as electronegativity.

At first sight their findings, based on the density-functional theory, that μ, the chemical potential, is the same for the entire system and equal for each orbital, is somewhat puzzling, if $-\mu$ can be identified with the electronegativity of the system as they suggest. This would mean the same electronegativity for the entire system; thus there will be no electro- or nucleophilic centers in the system in contradiction to chemical experience. Furthermore, if two subsystems, atoms or molecules, are considered together as one system, they will aquire the same electronegativity, no matter how strong they interact or even, if they don't interact at all. The same objection may be raised against our orbital definition in the case of delocalized orbitals; however, orbitals may be localized.

In order to gain insight into these problems, let us first consider a two-electron system, where it is possible for singlet states to explicate the density functional exactly. We may write the exact singlet-coupled two-electron wavefunction as

$$\Psi(1,2) = \sum_{i} \{\varphi_i^2\} c_i \tag{A20}$$

where

$$\{\varphi_i^2\} = \frac{1}{\sqrt{2}} \begin{vmatrix} \varphi_i\alpha(1) & \varphi_i\beta(1) \\ \varphi_i\alpha(2) & \varphi_i\beta(2) \end{vmatrix} \tag{A21}$$

constructed with orbitals φ_i determined in an MCSCF sense, i.e. natural orbitals, with Ψ normalized and the orbitals φ_i orthonormal. The charge density for such a function is

$$\varrho(1) = \sum_i 2\, c_i^* c_i\, \varphi_i^*(1)\varphi_i(1) \tag{A22}$$

with the orbital occupation number

$$n_i = 2\, c_i^* c_i \tag{A23}$$

We may without loss of generality assume the CI expansion coefficients to be real and obtain

$$c_i = \text{sign}(c_i)\sqrt{n_i/2} \tag{A24}$$

The expectation value of the energy of this system becomes

$$\begin{aligned} \langle E \rangle &= \langle \Psi | H | \Psi \rangle = \sum_{ij} c_i^* \langle \{\varphi_i^2\} | H | \{\varphi_j^2\} \rangle c_j \\ &= \sum_i c_i^2 (2\, h_{ii} + J_{ii}) + \sum_{i>j} 2\, c_i c_j K_{ij} \end{aligned} \tag{A24a}$$

with the conventional meaning of the one electron integrals h_{ii} and the Coulomb- and exchange integrals J_{ii} and K_{ij} respectively.

Using Eq. (A24) we may rewrite the energy as a function of the occupation numbers as

$$E(n) = \langle E \rangle = \sum_i n_i (h_{ii} + 1/2\, J_{ii}) + \sum_{i>j} \text{sign}(c_i c_j) \sqrt{n_i}\sqrt{n_j}\, K_{ij} \tag{A25}$$

For the ground state or also some other states where one of the c_i's, say c_0 is clearly dominant we can eliminate the sign $(c_i c_j)$, for we know then, if c_0 is positive all other c_i's will be negative, because K_{ij} is always positive. Thus we obtain for those states

$$E(n) = \sum_i n_i (h_{ii} + 1/2\, J_{ii}) - \sum_{j>0} \sqrt{n_0}\sqrt{n_j}\, K_{0j} + \sum_{i>j>0} \sqrt{n_i}\sqrt{n_j}\, K_{ij} \tag{A26}$$

This function may now be extremalized with the restrictive conditions

$$\sum_i n_i = \sum_i 2\,|c_i|^2 = N = 2$$

and (A27)

$$\langle \varphi_i | \varphi_j \rangle = \delta_{ij}$$

added with Lagrange multipliers. This yields an equation akin to Eq. (54) of Ref. 5.

$$\delta \left\{ E(n, \varphi) - \mu N(n) - \sum_{ij} \lambda_{ij} \langle \varphi_i | \varphi_j \rangle \right\} = 0 \tag{A28}$$

giving the stationary condition akin to Eq. (56) of Ref. 5

$$0 = \sum_i \delta n_i \left[\frac{\partial E(n,\varphi)}{\partial n_i} - \mu \right]$$

$$+ \sum_i \int \delta\varphi_i \left[\frac{\partial E(n,\varphi)}{\partial \varphi_i} - 2 \sum_j \lambda_{ij}\varphi_j \right] \quad \text{(A29)}$$

As this equation must hold for any δn_i we have

$$\left[\frac{\delta E(n,\varphi)}{\delta n_i} - \mu \right] = 0 \quad \text{for all i} \quad \text{(A30)}$$

Using (A26) we may write (A30) explicitly

$$h_{ii} + 1/2\, J_{ii} + \frac{1}{2\sqrt{n_i}} \left[\sum_{\substack{j \neq i \\ j>0}} \sqrt{n_j}\, K_{ij} - (1 - \delta_{i0})\sqrt{n_0}\, K_{i0} \right] - \mu = 0 \quad \text{(A31)}$$

Thus we can identify μ here as being

$$\mu = \frac{E}{2} \quad \text{(A32)}$$

or in general the value of the stationary-state energy divided by the occupation number. We conclude that the chemical potential, μ, is not $(\partial E/\partial N)$ and not $(\partial E/\partial n_i)$, and thus it cannot be identified with electronegativity.

What we have explicitly shown above for a two-electron system may also be demonstrated in general. To this end we use a general CI expansion

$$\Psi = \Sigma_I \Phi_I C_I \quad \text{(A33)}$$

with

$$\langle \Phi_I | \Phi_J \rangle = \delta_{IJ}$$

and extremalize the expectation value of the energy

$$\langle E \rangle = \langle \Psi | H | \Psi \rangle = \sum_{IJ} C_I^* H_{IJ} C_J \quad \text{(A34)}$$

under the restriction

$$N = \left\langle \Psi \right| \sum_i a_i^+ a_i \left| \Psi \right\rangle = N \sum_I |C_I|^2 \quad \text{(A35)}$$

which is equivalent to the standard CI normalization condition

$$\sum_I |C_I|^2 = 1 \quad \text{(A36)}$$

The variational equation akin to Eq. (7) of Ref. 5 is now

$$\delta\{\langle E\rangle - \mu N\} = 0 \tag{A37}$$

which, for a variation of the CI-coefficients, leads to the standard CI eigenvalue equation. Thus μ is to be identified with E/N, the stationary-state energy/electron.

Appendix B

Here we will give an explicit derivation of the set of linear equations, Eq. (25), obtained using the concept of electronegativity equalization in a bond. In a given molecule with N localized two-center, two-electron bonds we will have 2N atomic orbitals, φ_{A_i}, forming these bonds. Here the index A signifies the atom, and i is a running index numbering in some fixed order all the atomic orbitals engaged in bond formation. In the bond between the orbitals i and j with $i > j$ we will have due to electronegativity equalization, see Eq. (25).

$$\chi_i(r_i, q_i^0 + \Delta q_i) = \chi_j(r_j, q_j^0 + \Delta q_j) \tag{B1}$$

As we need to consider each bond constructed from two orbitals only once we may restrict ourselves to those equations with $i > j$ after an atomic orbital order has been fixed. From each such equation we obtain a single charge shift in the bond between the orbitals i and j, Δq_{ij}; the sign of the individual charge shifts into the orbitals i and j respectively is fixed by

$$\Delta q_{ij} = \mathrm{sign}(i - j)\Delta q_i = \mathrm{sign}(j - i)\Delta q_j \tag{B2}$$

With this specification of the sign of the individual charges in the orbitals we are in a position to specify Q_A, Eq. (17), and r_{Ai}, Eq. (18), uniquely.

$$Q_A = \sum_{k \varepsilon A} \mathrm{sign}(k - l_k)\Delta q_{[k,l_k]} \tag{B3}$$

Here l_k is the orbital forming a bond with the orbital k on atom A. The square bracketed index $[k, l_k]$ is to indicate that the index is either k,l_k or l_k,k such that the first index of this double index is the larger. With the notation thus specified we have

$$r_{Ai} = \sum_{\substack{k \neq i \\ \varepsilon A}} \mathrm{sign}(k - l_k)\Delta q_{[k,l_k]} \tag{B4}$$

Substituting this together with Eq. (24) into (B1) gives

$$\begin{aligned} & b_i^0 + 2\,c_i q_i^0 + 2\,c_i \Delta q_{ij} + \sum_{\substack{k \neq i \\ \varepsilon A}} \mathrm{sign}(k - l_k) b_i^1 \Delta q_{[k,l_k]} \\ & = b_j^0 + 2\,c_j \Delta q_{ij} + \sum_{\substack{k \neq j \\ \varepsilon B}} \mathrm{sign}(k - l_k) b_j^1 \Delta q_{[k,l_k]} \end{aligned} \tag{B5}$$

where A and B are the atoms of the orbitals i and j respectively. With the definition of the standard values of the orbital electronegativity

$$\chi_i^0 = \chi_i(r_i = 0, q_i^0) = b_i^0 + 2\, c_i q_i^0 \tag{B6}$$

we rewrite (B5) to obtain

$$2\,(c_i + c_j)\Delta q_{ij} + \sum_{\substack{k \neq i \\ \varepsilon A}} \operatorname{sign}(k - l_k) b_i^1 \Delta q_{[k,l_k]} + \sum_{\substack{k \neq j \\ \varepsilon B}} \operatorname{sign}(k - l_k) b_j^1 \Delta q_{[k,l_k]} = \chi_j^0 - \chi_i^0 \tag{B7}$$

This is the representative final equation desired. These equations constructed for each bond with $i > j$ form a set of N linear equations for the determination of the N unknowns Δq_{ij}.

Conclusion and Acknowledgements. We have given a rigorous quantum-mechanical definition of orbital electronegativity and demonstrated the efficacy of this concept in determining charge distributions and other properties of molecules. We have shown that the chemical potential μ of the density-theoretical formulation[5] should not be identified with electronegativity or orbital electronegativity. Discussions with J. Broad and P. Hamacher, but especially with P. Pfeifer have contributed significantly in clarifying these issues. We are grateful for their interest and insights as well as for the computational assistance of U. Welz.

Financial support of the "Fond der Chemischen Industrie" is gratefully acknowledged.

G. References

1. Pauling, L.: Proc. Natl. Acad. Sci. USA *84,* 414 (1932); J. Am. Chem. Soc. *54,* 3570 (1932); "The Nature of the Chemical Bond", Ithaca, N.Y.: M. P. Cornell 3rd Ed. (1960)
2. Hinze, J.: Fortschr. Chem. Forsch. *9,* 448 (1968)
   Wells, P. R.: Prog. Phys. Org. Chem. *6,* 111 (1968)
3. Malone, J. G.: J. Chem. Phys. *1,* 197 (1933)
   Walsh, A. D.: Proc. Roy. Soc. (London) Ser. A *207,* 13 (1951)
   Gordy, W.: Phys. Rev. *69,* 130 (1946); J. Chem. Phys. *14,* 305 (1946); J. Chem. Phys. *19,* 792 (1950)
   Gordy, W., Thomas, W. J. O.: J. Chem. Phys. *24,* 439 (1956)
   Dailey, B. P., Shoolerey, J. N.: J. Am. Chem. Soc. *77,* 3977 (1955)
   Shoolerey, J. N.: J. Chem. Phys. *21,* 1899 (1953)
   Heel, H., Zeil, W.: Z. Elektrochem. *64,* 962 (1960)
   Zeil, W., Burchert, H.: Z. Physik. Chem. (Frankfurt) *38,* 47 (1963)
   Jørgensen, C. K.: Mol. Phys. *6,* 43 (1963); Danski Kemi *45,* 113 (1964)
   Clifford, A. F.: J. Phys. Chem. *63,* 1227 (1959)
   Moffitt, W.: Proc. Roy. Soc. (London) Ser. A *202,* 534 (1950)
   Phillips, J. C.: J. Phys. Chem. Solids *34,* 1051 (1973)
   Miedema, A. R.: J. Less Common Met. *32,* 117 (1973)
   Miedema, A. R., De Boer, F. R., De Chatel, P. F.: J. Phys. F., Met. Phys. *3,* 1558 (1973)
4. Spiridanov, V. P., Tatevskii, V. M.: Zh. Fiz. Khim. *37,* 994, 1236, 1583, 1973, 2173 (1963)
   Tatevskii, N. M., Spiridanov, V. P.: Zh. Khim. *39,* 1284 (1964)
5. Parr, R. G., Donelly, R. A., Levy, M., Palke, W. E.: J. Chem. Phys. *68,* 3801 (1978)
6. Iczkowski, R. P., Margrave, J. L.: J. Am. Chem. Soc. *83,* 3547 (1961)
7. Hinze, J., Whitehead, M. A., Jaffé, H. H.: J. Am. Chem. Soc. *85,* 148 (1963)

8. Robles, J., Bartolotti, L. J.: J. Am. Chem. Soc. *106,* 3723 (1984)
Baroni, S., Tuncel, E.: J. Chem. Phys. *79,* 6140 (1983)
Boyd, R. J., Markus, G. E.: J. Chem. Phys. *75,* 5385 (1981)
Manoli, S., Whitehead, M. A.: J. Chem. Phys. *81,* 841 (1984)
Mullay, J.: J. Am. Chem. Soc. *106,* 5842 (1984)
Gázquez, J. L., Ortiz, E.: J. Chem. Phys. *81,* 2741 (1984)
Sen, K. D.: J. Phys. B: At. Mol. Phys. *16,* L 149 (1983)
9. Sanderson, R. T.: Science *114,* 670 (1951); "Chemical Bonds and Bond Energy", Acad. Press, 2nd Ed. (1976)
10. Gasteiger, J., Hutchings, M. G.: J. Am. Chem. Soc. *106,* 22 (1984)
Hutchings, M. G., Gasteiger, J.: Tetrahedron Lett. *24,* 2541 (1983)
Huheey, J. E.: J. Org. Chem. *31,* 2365 (1966)
Hanson, P.: J. Chem. Soc. Perkin Trans. II, 101 (1984)
Jardine, W. K., Langler, R. F., MacGregor, J. A.: Can. J. Chem. *60,* 2069 (1982)
Watson, R. E., Benett, L. H., Davenport, J. W.: Phys. Rev. B *27,* 6428 (1983)
Huheey, J. E.: J. Phys. Chem. *69,* 3284 (1965)
Gasteiger, J., Marsili, M.: Tetrahedron *36,* 3219 (1980)
11. Parr, R. G., Pearson, R. G.: J. Am. Chem. Soc. *105,* 7512 (1983)
12. Mulliken, R. S.: J. Chem. Phys. *2,* 782 (1934); J. Chem. Phys. *3,* 513 (1935)
13. Hinze, J., Jaffé, H. H.: J. Am. Chem. Soc. *84,* 540 (1962)
14. Hinze, J., Whitehead, M. A., Jaffé, H. H.: J. Am. Chem. Soc. *85,* 148 (1963)
15. Hinze, J., Jaffé, H. H.: Can. J. Chem. *41,* 1315 (1963)
16. Bergmann, D.: Diplomarbeit, Universität Bielefeld (1985)
17. Moore, C. E.: "Atomic Energy Levels", Natl. Bur. Stand., Vol. I–III (1959)
Hotop, H., Lineberger, W. C.: J. Phys. Chem. Ref. Data *4,* 539 (1975)
18. Polansky, O. E., Derflinger, G.: Theor. Chim. Acta (Berl.) *1,* 308 (1963)
Derflinger, G., Polansky, O. E.: Theor. Chim. Acta (Berl.) *1,* 316 (1963)
19. Dailey, P., Shoolerey, J. N.: J. Am. Chem. Soc. *77,* 3972 (1955)
20. Lias, S. G., Liebmann, J. F., Levin, R. D.: J. Phys. Chem. Ref. Data *13,* 695 (1984)
21. Gasteiger, J., Marsili, M.: Tetrahedron *36,* 3219 (1980)
Mortier, W. J., Van Gnechten, K., Gasteiger, J.: J. Am. Chem. Soc. *107,* 829 (1985)
22. Hohenberg, P., Kohn, W.: Phys. Rev. B *136,* 864 (1964)
23. Hinze, J.: J. Chem. Phys. *59,* 6424 (1973)
24. Hinze, J.: to be published in J. Chem. Phys.
25. Mariott, S., Topsom, R. D.: J. Mol. Struct. *89,* 83 (1982)
26. Schomaker, V., Stephenson, D. P.: J. Am. Chem. Soc. *63,* 37 (1941)
27. "Sadtler Standard ^{13}C-NMR-Spectra", Sadtler Research Laboratories, Philadelphia, P.A., USA (1976)

Author Index Volumes 1–66

Ahrland, S.: Factors Contributing to (b)-behaviour in Acceptors. Vol. 1, pp. 207–220.

Ahrland, S.: Thermodynamics of Complex Formation between Hard and Soft Acceptors and Donors. Vol. 5, pp. 118–149.

Ahrland, S.: Thermodynamics of the Stepwise Formation of Metal-Ion Complexes in Aqueous Solution. Vol. 15, pp. 167–188.

Allen, G. C., Warren, K. D.: The Electronic Spectra of the Hexafluoro Complexes of the First Transition Series. Vol. 9, pp. 49–138.

Allen, G. C., Warren, K. D.: The Electronic Spectra of the Hexafluoro Complexes of the Second and Third Transition Series. Vol. 19, pp. 105–165.

Alonso, J. A., Balbás, L. C.: Simple Density Functional Theory of the Electronegativity and Other Related Properties of Atoms and Ions. Vol. 66, pp. 41–78.

Ardon, M., Bino, A.: A New Aspect of Hydrolysis of Metal Ions: The Hydrogen-Oxide Bridging Ligand ($H_3O_2^-$). Vol. 65, pp. 1–28.

Averill, B. A.: Fe–S and Mo–Fe–S Clusters as Models for the Active Site of Nitrogenase. Vol. 53, pp. 57–101.

Babel, D.: Structural Chemistry of Octahedral Fluorocomplexes of the Transition Elements. Vol. 3, pp. 1–87.

Bacci, M.: The Role of Vibronic Coupling in the Interpretation of Spectroscopic and Structural Properties of Biomolecules. Vol. 55, pp. 67–99.

Baker, E. C., Halstead, G.W., Raymond, K. N.: The Structure and Bonding of 4*f* and 5*f* Series Organometallic Compounds. Vol. 25, pp. 21–66.

Balsenc, L. R.: Sulfur Interaction with Surfaces and Interfaces Studied by Auger Electron Spectrometry. Vol. 39, pp. 83–114.

Banci, L., Bencini, A., Benelli, C., Gatteschi, D., Zanchini, C.: Spectral-Structural Correlations in High-Spin Cobalt(II) Complexes. Vol. 52, pp. 37–86.

Bartolotti, L. J.: Absolute Electronegativities as Determined from Kohn-Sham Theory. Vol. 66, pp. 27–40.

Baughan, E. C.: Structural Radii, Electron-cloud Radii, Ionic Radii and Solvation. Vol. 15, pp. 53–71.

Bayer, E., Schretzmann, P.: Reversible Oxygenierung von Metallkomplexen. Vol. 2, pp. 181–250.

Bearden, A. J., Dunham, W. R.: Iron Electronic Configurations in Proteins: Studies by Mössbauer Spectroscopy. Vol. 8, pp. 1–52.

Bergmann, D., Hinze, J.: Electronegativity and Charge Distribution. Vol. 66, pp. 145–190.

Bertini, I., Luchinat, C., Scozzafava, A.: Carbonic Anhydrase: An Insight into the Zinc Binding Site and into the Active Cavity Through Metal Substitution. Vol. 48, pp. 45–91.

Blasse, G.: The Influence of Charge-Transfer and Rydberg States on the Luminescence Properties of Lanthanides and Actinides. Vol. 26, pp. 43–79.

Blasse, G.: The Luminescence of Closed-Shell Transition Metal-Complexes. New Developments. Vol. 42, pp. 1–41.

Blauer, G.: Optical Activity of Conjugated Proteins. Vol. 18, pp. 69–129.

Bleijenberg, K. C.: Luminescence Properties of Uranate Centres in Solids. Vol. 42, pp. 97–128.

Boeyens, J. C. A.: Molecular Mechanics and the Structure Hypothesis. Vol. 63, pp. 65–101.

Bonnelle, C.: Band and Localized States in Metallic Thorium, Uranium and Plutonium, and in Some Compounds, Studied by X-Ray Spectroscopy. Vol. 31, pp. 23–48.

Bradshaw, A. M., Cederbaum, L. S., Domcke, W.: Ultraviolet Photoelectron Spectroscopy of Gases Adsorbed on Metal Surfaces. Vol. 24, pp. 133–170.

Braterman, P. S.: Spectra and Bonding in Metal Carbonyls. Part A: Bonding. Vol. 10, pp. 57–86.

Braterman, P. S.: Spectra and Bonding in Metal Carbonyls. Part B: Spectra and Their Interpretation. Vol. 26, pp. 1–42.

Bray, R. C., Swann, J. C.: Molybdenum-Containing Enzymes. Vol. 11, pp. 107–144.

Brooks, M. S. S.: The Theory of 5f Bonding in Actinide Solids. Vol. 59/60, pp. 263–293.

van Bronswyk, W.: The Application of Nuclear Quadrupole Resonance Spectroscopy to the Study of Transition Metal Compounds. Vol. 7, pp. 87–113.

Buchanan, B. B.: The Chemistry and Function of Ferredoxin. Vol. 1, pp. 109–148.

Buchler, J. W., Kokisch, W., Smith, P. D.: Cis, Trans, and *Metal* Effects in Transition Metal Porphyrins. Vol. 34, pp. 79–134.

Bulman, R. A.: Chemistry of Plutonium and the Transuranics in the Biosphere. Vol. 34, pp. 39–77.

Burdett, J. K.: The Shapes of Main-Group Molecules; A Simple Semi-Quantitative Molecular Orbital Approach. Vol. 31, pp. 67–105.
Burdett, J. K.: Some Structural Problems Examined Using the Method of Moments. Vol. 65, pp. 29–90.
Campagna, M., Wertheim, G. K., Bucher, E.: Spectroscopy of Homogeneous Mixed Valence Rare Earth Compounds. Vol. 30, pp. 99–140.
Chasteen, N. D.: The Biochemistry of Vanadium, Vol. 53, pp. 103–136.
Cheh, A. M., Neilands, J. P.: The δ-Aminolevulinate Dehydratases: Molecular and Environmental Properties. Vol. 29, pp. 123–169.
Ciampolini, M.: Spectra of 3d Five-Coordinate Complexes. Vol. 6, pp. 52–93.
Chimiak, A., Neilands, J. B.: Lysine Analogues of Siderophores. Vol. 58, pp. 89–96.
Clack, D. W., Warren, K. D.: Metal-Ligand Bonding in 3d Sandwich Complexes, Vol. 39, pp. 1–41.
Clark, R. J. H., Stewart, B.: The Resonance Raman Effect. Review of the Theory and of Applications in Inorganic Chemistry. Vol. 36, pp. 1–80.
Clarke, M. J., Fackler, P. H.: The Chemistry of Technetium: Toward Improved Diagnostic Agents. Vol. 50, pp. 57–78.
Cohen, I. A.: Metal-Metal Interactions in Metalloporphyrins, Metalloproteins and Metalloenzymes. Vol. 40, pp. 1–37.
Connett, P. H., Wetterhahn, K. E.: Metabolism of the Carcinogen Chromate by Cellular Constitutents. Vol. 54, pp. 93–124.
Cook, D. B.: The Approximate Calculation of Molecular Electronic Structures as a Theory of Valence. Vol. 35, pp. 37–86.
Cotton, F. A., Walton, R. A.: Metal-Metal Multiple Bonds in Dinuclear Clusters. Vol. 62, pp. 1–49.
Cox, P. A.: Fractional Parentage Methods for Ionisation of Open Shells of *d* and *f* Electrons. Vol. 24, pp. 59–81.
Crichton, R. R.: Ferritin. Vol. 17, pp. 67–134.
Daul, C., Schläpfer, C. W., von Zelewsky, A.: The Electronic Structure of Cobalt(II) Complexes with Schiff Bases and Related Ligands. Vol. 36, pp. 129–171.
Dehnicke, K., Shihada, A.-F.: Structural and Bonding Aspects in Phosphorus Chemistry-Inorganic Derivates of Oxohalogeno Phosphoric Acids. Vol. 28, pp. 51–82.
Dobiáš, B.: Surfactant Adsorption on Minerals Related to Flotation. Vol. 56, pp. 91–147.
Doughty, M. J., Diehn, B.: Flavins as Photoreceptor Pigments for Behavioral Responses. Vol. 41, pp. 45–70.
Drago, R. S.: Quantitative Evaluation and Prediction of Donor-Acceptor Interactions. Vol. 15, pp. 73–139.
Duffy, J. A.: Optical Electronegativity and Nephelauxetic Effect in Oxide Systems. Vol. 32, pp. 147–166.
Dunn, M. F.: Mechanisms of Zinc Ion Catalysis in Small Molecules and Enzymes. Vol. 23, pp. 61–122.
Emsley, E.: The Composition, Structure and Hydrogen Bonding of the β-Deketones. Vol. 57, pp. 147–191.
Englman, R.: Vibrations in Interaction with Impurities. Vol. 43, pp. 113–158.
Epstein, I. R., Kustin, K.: Design of Inorganic Chemical Oscillators. Vol. 56, pp. 1–33.
Ermer, O.: Calculations of Molecular Properties Using Force Fields. Applications in Organic Chemistry. Vol. 27, pp. 161–211.
Ernst, R. D.: Structure and Bonding in Metal-Pentadienyl and Related Compounds. Vol. 57, pp. 1–53.
Erskine, R. W., Field, B. O.: Reversible Oxygenation. Vol. 28, pp. 1–50.
Fajans, K.: Degrees of Polarity and Mutual Polarization of Ions in the Molecules of Alkali Fluorides, SrO, and BaO. Vol. 3, pp. 88–105.
Fee, J. A.: Copper Proteins – Systems Containing the "Blue" Copper Center. Vol. 23, pp. 1–60.
Feeney, R. E., Komatsu, S. K.: The Transferrins. Vol. 1, pp. 149–206.
Felsche, J.: The Crystal Chemistry of the Rare-Earth Silicates. Vol. 13, pp. 99–197.
Ferreira, R.: Paradoxical Violations of Koopmans' Theorem, with Special Reference to the 3d Transition Elements and the Lanthanides. Vol. 31, pp. 1–21.
Fidelis, I. K., Mioduski, T.: Double-Double Effect in the Inner Transition Elements. Vol. 47, pp. 27–51.
Fournier, J. M.: Magnetic Properties of Actinide Solids. Vol. 59/60, pp. 127–196.
Fournier, J. M., Manes, L.: Actinide Solids. 5f Dependence of Physical Properties. Vol. 59/60, pp. 1–56.

Fraga, S., Valdemoro, C.: Quantum Chemical Studies on the Submolecular Structure of the Nucleic Acids. Vol. 4, pp. 1–62.

Fraústo da Silva, J. J. R., Williams, R. J. P.: The Uptake of Elements by Biological Systems. Vol. 29, pp. 67–121.

Fricke, B.: Superheavy Elements. Vol. 21, pp. 89–144.

Fuhrhop, J.-H.: The Oxidation States and Reversible Redox Reactions of Metalloporphyrins. Vol. 18, pp. 1–67.

Furlani, C., Cauletti, C.: He(I) Photoelectron Spectra of *d*-metal Compounds. Vol. 35, pp. 119–169.

Gázques, J. L., Vela, A., Galván, M.: Fukui Function, Electronegativity and Hardness in the Kohn-Sham Theory. Vol. 66, pp. 79–98.

Gerloch, M., Harding, J. H., Woolley, R. G.: The Context and Application of Ligand Field Theory. Vol. 46, pp. 1–46.

Gillard, R. D., Mitchell, P. R.: The Absolute Configuration of Transition Metal Complexes. Vol. 7, pp. 46–86.

Gleitzer, C., Goodenough, J. B.: Mixed-Valence Iron Oxides. Vol. 61, pp. 1–76.

Gliemann, G., Yersin, H.: Spectroscopic Properties of the Quasi One-Dimensional Tetracyanoplatinate(II) Compounds. Vol. 62, pp. 87–153.

Golovina, A. P., Zorov, N. B., Runov, V. K.: Chemical Luminescence Analysis of Inorganic Substances. Vol. 47, pp. 53–119.

Green, J. C.: Gas Phase Photoelectron Spectra of *d*- and *f*-Block Organometallic Compounds. Vol. 43, pp. 37–112.

Grenier, J. C., Pouchard, M., Hagenmuller, P.: Vacancy Ordering in Oxygen-Deficient Perovskite-Related Ferrities. Vol. 47, pp. 1–25.

Griffith, J. S.: On the General Theory of Magnetic Susceptibilities of Polynuclear Transitionmetal Compounds. Vol. 10, pp. 87–126.

Gubelmann, M. H., Williams, A. F.: The Structure and Reactivity of Dioxygen Complexes of the Transition Metals. Vol. 55, pp. 1–65.

Guilard, R., Lecomte, C., Kadish, K. M.: Synthesis, Electrochemistry, and Structural Properties of Porphyrins with Metal-Carbon Single Bonds and Metal-Metal Bonds. Vol. 64, pp. 205–268.

Gütlich, P.: Spin Crossover in Iron(II)-Complexes. Vol. 44, pp. 83–195.

Gutmann, V., Mayer, U.: Thermochemistry of the Chemical Bond. Vol. 10, pp. 127–151.

Gutmann, V., Mayer, U.: Redox Properties: Changes Effected by Coordination. Vol. 15, pp. 141–166.

Gutmann, V., Mayer, H.: Application of the Functional Approach to Bond Variations under Pressure. Vol. 31, pp. 49–66.

Hall, D. I., Ling, J. H., Nyholm, R. S.: Metal Complexes of Chelating Olefin-Group V Ligands. Vol. 15, pp. 3–51.

Harnung, S. E., Schäffer, C. E.: Phase-fixed 3-Γ Symbols and Coupling Coefficients for the Point Groups. Vol. 12, pp. 201–255.

Harnung, S. E., Schäffer, C. E.: Real Irreducible Tensorial Sets and their Application to the Ligand-Field Theory. Vol. 12, pp. 257–295.

Hathaway, B. J.: The Evidence for "Out-of-the-Plane" Bonding in Axial Complexes of the Copper(II) Ion. Vol. 14, pp. 49–67.

Hathaway, B. J.: A New Look at the Stereochemistry and Electronic Properties of Complexes of the Copper(II) Ion. Vol. 57, pp. 55–118.

Hellner, E. E.: The Frameworks (Bauverbände) of the Cubic Structure Types. Vol. 37, pp. 61–140.

von Herigonte, P.: Electron Correlation in the Seventies. Vol. 12, pp. 1–47.

Hemmerich, P., Michel, H., Schug, C., Massey, V.: Scope and Limitation of Single Electron Transfer in Biology. Vol. 48, pp. 93–124.

Hider, R. C.: Siderophores Mediated Absorption of Iron. Vol. 58, pp. 25–88.

Hill, H. A. O., Röder, A., Williams, R. J. P.: The Chemical Nature and Reactivity of Cytochrome P-450. Vol. 8, pp. 123–151.

Hogenkamp, H. P. C., Sando, G. N.: The Enzymatic Reduction of Ribonucleotides. Vol. 20, pp. 23–58.

Hoffmann, D. K., Ruedenberg, K., Verkade, J. G.: Molecular Orbital Bonding Concepts in Polyatomic Molecules – A Novel Pictorial Approach. Vol. 33, pp. 57–96.

Hubert, S., Hussonnois, M., Guillaumont, R.: Measurement of Complexing Constants by Radiochemical Methods. Vol. 34, pp. 1–18.

Hudson, R. F.: Displacement Reactions and the Concept of Soft and Hard Acids and Bases. Vol. 1, pp. 221–223.

Hulliger, F.: Crystal Chemistry of Chalcogenides and Pnictides of the Transition Elements. Vol. 4, pp. 83–229.

Ibers, J. A., Pace, L. J., Martinsen, J., Hoffman, B. M.: Stacked Metal Complexes: Structures and Properties. Vol. 50, pp. 1–55.

Iqbal, Z.: Intra- und Inter-Molecular Bonding and Structure of Inorganic Pseudohalides with Triatomic Groupings. Vol. 10, pp. 25–55.

Izatt, R. M., Eatough, D. J., Christensen, J. J.: Thermodynamics of Cation-Macrocyclic Compound Interaction. Vol. 16, pp. 161–189.

Jain, V. K., Bohra, R., Mehrotra, R. C.: Structure and Bonding in Organic Derivatives of Antimony(V). Vol. 52, pp. 147–196.

Jerome-Lerutte, S.: Vibrational Spectra and Structural Properties of Complex Tetracyanides of Platinum, Palladium and Nickel. Vol. 10, pp. 153–166.

Jørgensen, C. K.: Electric Polarizability, Innocent Ligands and Spectroscopic Oxidation States. Vol. 1, pp. 234–248.

Jørgensen, C. K.: Recent Progress in Ligand Field Theory. Vol. 1, pp. 3–31.

Jørgensen, C. K.: Relations between Softness, Covalent Bonding, Ionicity and Electric Polarizability. Vol. 3, pp. 106–115.

Jørgensen, C. K.: Valence-Shell Expansion Studied by Ultra-violet Spectroscopy. Vol. 6, pp. 94–115.

Jørgensen, C. K.: The Inner Mechanism of Rare Earths Elucidated by Photo-Electron Spectra. Vol. 13, pp. 199–253.

Jørgensen, C. K.: Partly Filled Shells Constituting Anti-bonding Orbitals with Higher Ionization Energy than their Bonding Counterparts. Vol. 22, pp. 49–81.

Jørgensen, C. K.: Photo-electron Spectra of Non-metallic Solids and Consequences for Quantum Chemistry. Vol. 24, pp. 1–58.

Jørgensen, C. K.: Narrow Band Thermoluminescence (Candoluminescence) of Rare Earths in Auer Mantles. Vol. 25, pp. 1–20.

Jørgensen, C. K.: Deep-lying Valence Orbitals and Problems of Degeneracy and Intensities in Photoelectron Spectra. Vol. 30, pp. 141–192.

Jørgensen, C. K.: Predictable Quarkonium Chemistry. Vol. 34, pp. 19–38.

Jørgensen, C. K.: The Conditions for Total Symmetry Stabilizing Molecules, Atoms, Nuclei and Hadrons. Vol. 43, pp. 1–36.

Jørgensen, C. K., Reisfeld, R.: Uranyl Photophysics. Vol. 50, pp. 121–171.

O'Keeffe, M., Hyde, B. G.: An Alternative Approach to Non-Molecular Crystal Structures with Emphasis on the Arrangements of Cations. Vol. 61, pp. 77–144.

Kimura, T.: Biochemical Aspects of Iron Sulfur Linkage in None-Heme Iron Protein, with Special Reference to "Adrenodoxin". Vol. 5, pp. 1–40.

Kitagawa, T., Ozaki, Y.: Infrared and Raman Spectra of Metalloporphyrins. Vol. 64, pp. 71–114.

Kiwi, J., Kalyanasundaram, K., Grätzel, M.: Visible Light Induced Cleavage of Water into Hydrogen and Oxygen in Colloidal and Microheterogeneous Systems. Vol. 49, pp. 37–125.

Kjekshus, A., Rakke, T.: Considerations on the Valence Concept. Vol. 19, pp. 45–83.

Kjekshus, A., Rakke, T.: Geometrical Considerations on the Marcasite Type Structure. Vol. 19, pp. 85–104.

König, E.: The Nephelauxetic Effect. Calculation and Accuracy of the Interelectronic Repulsion Parameters I. Cubic High-Spin d^2, d^3, d^7 and d^8 Systems. Vol. 9, pp. 175–212.

Koppikar, D. K., Sivapullaiah, P. V., Ramakrishnan, L., Soundararajan, S.: Complexes of the Lanthanides with Neutral Oxygen Donor Ligands. Vol. 34, pp. 135–213.

Krumholz, P.: Iron(II) Diimine and Related Complexes. Vol. 9, pp. 139–174.

Kustin, K., McLeod, G. C., Gilbert, T. R., Briggs, LeB. R., 4th.: Vanadium and Other Metal Ions in the Physiological Ecology of Marine Organisms. Vol. 53, pp. 137–158.

Labarre, J. F.: Conformational Analysis in Inorganic Chemistry: Semi-Empirical Quantum Calculation vs. Experiment. Vol. 35, pp. 1–35.

Lammers, M., Follmann, H.: The Ribonucleotide Reductases: A Unique Group of Metalloenzymes Essential for Cell Proliferation. Vol. 54, pp. 27–91.

Lehn, J.-M.: Design of Organic Complexing Agents. Strategies towards Properties. Vol. 16, pp. 1–69.

Linarès, C., Louat, A., Blanchard, M.: Rare-Earth Oxygen Bonding in the $LnMO_4$Xenotime Structure. Vol. 33, pp. 179–207.

Lindskog, S.: Cobalt(II) in Metalloenzymes. A Reporter of Structure-Function Relations. Vol. 8, pp. 153–196.

Liu, A., Neilands, J. B.: Mutational Analysis of Rhodotorulic Acid Synthesis in *Rhodotorula pilimanae*. Vol. 58, pp. 97–106.

Livorness, J., Smith, T.: The Role of Manganese in Photosynthesis. Vol. 48, pp. 1–44.

Llinás, M.: Metal-Polypeptide Interactions: The Conformational State of Iron Proteins. Vol. 17, pp. 135–220.

Lucken, E. A. C.: Valence-Shell Expansion Studied by Radio-Frequency Spectroscopy. Vol. 6, pp. 1–29.

Ludi, A., Güdel, H. U.: Structural Chemistry of Polynuclear Transition Metal Cyanides. Vol. 14, pp. 1–21.

Maggiora, G. M., Ingraham, L. L.: Chlorophyll Triplet States. Vol. 2, pp. 126–159.

Magyar, B.: Salzebullioskopie III. Vol. 14, pp. 111–140.

Makovicky, E., Hyde, B. G.: Non-Commensurate (Misfit) Layer Structures. Vol. 46, pp. 101–170.

Manes, L., Benedict, U.: Structural and Thermodynamic Properties of Actinide Solids and Their Relation to Bonding. Vol. 59/60, pp. 75–125.

Mann, S.: Mineralization in Biological Systems. Vol. 54, pp. 125–174.

Mason, S. F.: The Ligand Polarization Model for the Spectra of Metal Complexes: The Dynamic Coupling Transition Probabilities. Vol. 39, pp. 43–81.

Mathey, F., Fischer, J., Nelson, J. H.: Complexing Modes of the Phosphole Moiety. Vol. 55, pp. 153–201.

Mayer, U., Gutmann, V.: Phenomenological Approach to Cation-Solvent Interactions. Vol. 12, pp. 113–140.

Mildvan, A. S., Grisham, C. M.: The Role of Divalent Cations in the Mechanism of Enzyme Catalyzed Phosphoryl and Nucleotidyl. Vol. 20, pp. 1–21.

Mingos, D. M. P., Hawes, J. C.: Complementary Spherical Electron Density Model. Vol. 63, pp. 1–63.

Moreau-Colin, M. L.: Electronic Spectra and Structural Properties of Complex Tetracyanides of Platinum, Palladium and Nickel. Vol. 10, pp. 167–190.

Morgan, B., Dolphin, D.: Synthesis and Structure of Biometic Porphyrins. Vol. 64, pp. 115–204.

Morris, D. F. C.: Ionic Radii and Enthalpies of Hydration of Ions. Vol. 4, pp. 63–82.

Morris, D. F. C.: An Appendix to Structure and Bonding. Vol. 4 (1968). Vol. 6, pp. 157–159.

Mortier, J. W.: Electronegativity Equalization and its Applications. Vol. 66, pp. 125–143.

Müller, A., Baran, E. J., Carter, R. O.: Vibrational Spectra of Oxo-, Thio-, and Selenometallates of Transition Elements in the Solid State. Vol. 26, pp. 81–139.

Müller, A., Diemann, E., Jørgensen, C. K.: Electronic Spectra of Tetrahedral Oxo, Thio and Seleno Complexes Formed by Elements of the Beginning of the Transition Groups. Vol. 14, pp. 23–47.

Müller, U.: Strukturchemie der Azide. Vol. 14, pp. 141–172.

Müller, W., Spirlet, J.-C.: The Preparation of High Purity Actinide Metals and Compounds. Vol. 59/60, pp. 57–73.

Mullay, J. J.: Estimation of Atomic and Group Electronegativities. Vol. 66, pp. 1–25.

Murrell, J. N.: The Potential Energy Surfaces of Polyatomic Molecules. Vol. 32, pp. 93–146.

Naegele, J. R., Ghijsen, J.: Localization and Hybridization of 5f States in the Metallic and Ionic Bond as Investigated by Photoelectron Spectroscopy. Vol. 59/60, pp. 197–262.

Nag, K., Bose, S. N.: Chemistry of Tetra- and Pentavalent Chromium. Vol. 63, pp. 153–197.

Neilands, J. B.: Naturally Occurring Non-porphyrin Iron Compounds. Vol. 1, pp. 59–108.

Neilands, J. B.: Evolution of Biological Iron Binding Centers. Vol. 11, pp. 145–170.

Neilands, J. B.: Methodology of Siderophores. Vol. 58, pp. 1–24.

Nieboer, E.: The Lanthanide Ions as Structural Probes in Biological and Model Systems. Vol. 22, pp. 1–47.

Novack, A.: Hydrogen Bonding in Solids. Correlation of Spectroscopic and Christallographic Data. Vol. 18, pp. 177–216.

Nultsch, W., Häder, D.-P.: Light Perception and Sensory Transduction in Photosynthetic Prokaryotes. Vol. 41, pp. 111–139.

Odom, J. D.: Selenium Biochemistry. Chemical and Physical Studies. Vol. 54, pp. 1–26.

Oelkrug, D.: Absorption Spectra and Ligand Field Parameters of Tetragonal 3*d*-Transition Metal Fluorides. Vol. 9, pp. 1–26.

Oosterhuis, W. T.: The Electronic State of Iron in Some Natural Iron Compounds: Determination by Mössbauer and ESR Spectroscopy. Vol. 20, pp. 59–99.

Orchin, M., Bollinger, D. M.: Hydrogen-Deuterium Exchange in Aromatic Compounds. Vol. 23, pp. 167–193.

Peacock, R. D.: The Intensities of Lanthanide $f \longleftrightarrow f$ Transitions. Vol. 22, pp. 83–122.

Penneman, R. A., Ryan, R. R., Rosenzweig, A.: Structural Systematics in Actinide Fluoride Complexes. Vol. 13, pp. 1–52.

Powell, R. C., Blasse, G.: Energy Transfer in Concentrated Systems. Vol. 42, pp. 43–96.

Que, Jr., L.: Non-Heme Iron Dioxygenases. Structure and Mechanism. Vol. 40, pp. 39–72.

Ramakrishna, V. V., Patil, S. K.: Synergic Extraction of Actinides. Vol. 56, pp. 35–90.

Raymond, K. N., Smith, W. L.: Actinide-Specific Sequestering Agents and Decontamination Applications. Vol. 43, pp. 159–186.

Reinen, D.: Ligand-Field Spectroscopy and Chemical Bonding in Cr^{3+}-Containing Oxidic Solids. Vol. 6, pp. 30–51.

Reinen, D.: Kationenverteilung zweiwertiger $3d^n$-Ionen in oxidischen Spinell-, Granat- und anderen Strukturen. Vol. 7, pp. 114–154.

Reinen, D., Friebel, C.: Local and Cooperative Jahn-Teller Interactions in Model Structures. Spectroscopic and Structural Evidence. Vol. 37, pp. 1–60.

Reisfeld, R.: Spectra and Energy Transfer of Rare Earths in Inorganic Glasses. Vol. 13, pp. 53–98.

Reisfeld, R.: Radiative and Non-Radiative Transitions of Rare Earth Ions in Glasses. Vol. 22, pp. 123–175.

Reisfeld, R.: Excited States and Energy Transfer from Donor Cations to Rare Earths in the Condensed Phase. Vol. 30, pp. 65–97.

Reisfeld, R., Jørgensen, C. K.: Luminescent Solar Concentrators for Energy Conversion. Vol. 49, pp. 1–36.

Russo, V. E. A., Galland, P.: Sensory Physiology of *Phycomyces Blakesleeanus*. Vol. 41, pp. 71–110.

Rüdiger, W.: Phytochrome, a Light Receptor of Plant Photomorphogenesis. Vol. 40, pp. 101–140.

Ryan, R. R., Kubas, G. J., Moody, D. C., Eller, P. G.: Structure and Bonding of Transition Metal-Sulfur Dioxide Complexes. Vol. 46, pp. 47–100.

Sadler, P. J.: The Biological Chemistry of Gold: A Metallo-Drug and Heavy-Atom Label with Variable Valency. Vol. 29, pp. 171–214.

Schäffer, C. E.: A Perturbation Representation of Weak Covalent Bonding. Vol. 5, pp. 68–95.

Schäffer, C. E.: Two Symmetry Parameterizations of the Angular-Overlap Model of the Ligand-Field. Relation to the Crystal-Field Model. Vol. 14, pp. 69–110.

Scheidt, W. R., Lee, Y. J.: Recent Advances in the Stereochemistry of Metallotetrapyrroles. Vol. 64, pp. 1–70.

Schmid, G.: Developments in Transition Metal Cluster Chemistry. The Way to Large Clusters. Vol. 62, pp. 51–85.

Schmidt, P. C.: Electronic Structure of Intermetallic B 32 Type Zintl Phases. Vol. 65, pp. 91–133.

Schneider, W.: Kinetics and Mechanism of Metalloporphyrin Formation. Vol. 23, pp. 123–166.

Schubert, K.: The Two-Correlations Model, a Valence Model for Metallic Phases. Vol. 33, pp. 139–177.

Schutte, C. J. H.: The Ab-Initio Calculation of Molecular Vibrational Frequencies and Force Constants. Vol. 9, pp. 213–263.

Schweiger, A.: Electron Nuclear Double Resonance of Transition Metal Complexes with Organic Ligands. Vol. 51, pp. 1–122.

Sen, K. D., Böhm, M. C., Schmidt, P. C.: Electronegativity of Atoms and Molecular Fragments. Vol. 66, pp. 99–123.

Shamir, J.: Polyhalogen Cations. Vol. 37, pp. 141–210.

Shannon, R. D., Vincent, H.: Relationship between Covalency, Interatomic Distances, and Magnetic Properties in Halides and Chalcogenides. Vol. 19, pp. 1–43.

Shriver, D. F.: The Ambident Nature of Cyanide. Vol. 1, pp. 32–58.

Siegel, F. L.: Calcium-Binding Proteins. Vol. 17, pp. 221–268.

Simon, A.: Structure and Bonding with Alkali Metal Suboxides. Vol. 36, pp. 81–127.

Simon, W., Morf, W. E., Meier, P. Ch.: Specificity for Alkali and Alkaline Earth Cations of Synthetic and Natural Organic Complexing Agents in Membranes. Vol. 16, pp. 113–160.

Simonetta, M., Gavezzotti, A.: Extended Hückel Investigation of Reaction Mechanisms. Vol. 27, pp. 1–43.

Sinha, S. P.: Structure and Bonding in Highly Coordinated Lanthanide Complexes. Vol. 25, pp. 67–147.

Sinha, S. P.: A Systematic Correlation of the Properties of the f-Transition Metal Ions. Vol. 30, pp. 1–64.
Schmidt, W.: Physiological Bluelight Reception. Vol. 41, pp. 1–44.
Smith, D. W.: Ligand Field Splittings in Copper(II) Compounds. Vol. 12, pp. 49–112.
Smith, D. W., Williams, R. J. P.: The Spectra of Ferric Haems and Haemoproteins, Vol. 7, pp. 1–45.
Smith, D. W.: Applications of the Angular Overlap Model. Vol. 35, pp. 87–118.
Solomon, E. I., Penfield, K. W., Wilcox, D. E.: Active Sites in Copper Proteins. An Electric Structure Overview. Vol. 53, pp. 1–56.
Somorjai, G. A., Van Hove, M. A.: Adsorbed Monolayers on Solid Surfaces. Vol. 38, pp. 1–140.
Speakman, J. C.: Acid Salts of Carboxylic Acids, Crystals with some "Very Short" Hydrogen Bonds. Vol. 12, pp. 141–199.
Spiro, G., Saltman, P.: Polynuclear Complexes of Iron and their Biological Implications. Vol. 6, pp. 116–156.
Strohmeier, W.: Problem and Modell der homogenen Katalyse. Vol. 5, pp. 96–117.
Sugiura, Y., Nomoto, K.: Phytosiderophores – Structures and Properties of Mugineic Acids and Their Metal Complexes. Vol. 58, pp. 107–135.
Tam, S.-C., Williams, R. J. P.: Electrostatics and Biological Systems. Vol. 63, pp. 103–151.
Teller, R., Bau, R. G.: Crystallographic Studies of Transition Metal Hydride Complexes. Vol. 44, pp. 1–82.
Thompson, D. W.: Structure and Bonding in Inorganic Derivates of β-Diketones. Vol. 9, pp. 27–47.
Thomson, A. J., Williams, R. J. P., Reslova, S.: The Chemistry of Complexes Related to *cis*-$Pt(NH_3)_2Cl_2$. An Anti-Tumor Drug. Vol. 11, pp. 1–46.
Tofield, B. C.: The Study of Covalency by Magnetic Neutron Scattering. Vol. 21, pp. 1–87.
Trautwein, A.: Mössbauer-Spectroscopy on Heme Proteins. Vol. 20, pp. 101–167.
Tressaud, A., Dance, J.-M.: Relationships Between Structure and Low-Dimensional Magnetism in Fluorides. Vol. 52, pp. 87–146.
Tributsch, H.: Photoelectrochemical Energy Conversion Involving Transition Metal d-States and Intercalation of Layer Compounds. Vol. 49, pp. 127–175.
Truter, M. R.: Structures of Organic Complexes with Alkali Metal Ions. Vol. 16, pp. 71–111.
Umezawa, H., Takita, T.: The Bleomycins: Antitumor Copper-Binding Antibiotics. Vol. 40, pp. 73–99.
Vahrenkamp, H.: Recent Results in the Chemistry of Transition Metal Clusters with Organic Ligands. Vol. 32, pp. 1–56.
Valach, F., Koreň, B., Sivý, P., Melník, M.: Crystal Structure Non-Rigidity of Central Atoms for Mn(II), Fe(II), Fe(III), Co(II), Co(III), Ni(II), Cu(II) and Zn(II) Complexes. Vol. 55, pp. 101–151.
Wallace, W. E., Sankar, S. G., Rao, V. U. S.: Field Effects in Rare-Earth Intermetallic Compounds. Vol. 33, pp. 1–55.
Warren, K. D.: Ligand Field Theory of Metal Sandwich Complexes. Vol. 27, pp. 45–159.
Warren, K. D.: Ligand Field Theory of f-Orbital Sandwich Complexes. Vol. 33, pp. 97–137.
Warren, K. D.: Calculations of the Jahn-Teller Coupling Costants for d^x Systems in Octahedral Symmetry via the Angular Overlap Model. Vol. 57, pp. 119–145.
Watson, R. E., Perlman, M. L.: X-Ray Photoelectron Spectroscopy. Application to Metals and Alloys. Vol. 24, pp. 83–132.
Weakley, T. J. R.: Some Aspects of the Heteropolymolybdates and Heteropolytungstates. Vol. 18, pp. 131–176.
Wendin, G.: Breakdown of the One-Electron Pictures in Photoelectron Spectra. Vol. 45, pp. 1–130.
Weissbluth, M.: The Physics of Hemoglobin. Vol. 2, pp. 1–125.
Weser, U.: Chemistry and Structure of some Borate Polyol Compounds. Vol. 2, pp. 160–180.
Weser, U.: Reaction of some Transition Metals with Nucleic Acids and their Constituents. Vol. 5, pp. 41–67.
Weser, U.: Structural Aspects and Biochemical Function of Erythrocuprein. Vol. 17, pp. 1–65.
Weser, U.: Redox Reactions of Sulphur-Containing Amino-Acid Residues in Proteins and Metalloproteins, an XPS-Study. Vol. 61, pp. 145–160.
Willemse, J., Cras, J. A., Steggerda, J. J., Keijzers, C. P.: Dithiocarbamates of Transition Group Elements in "Unusual" Oxidation State. Vol. 28, pp. 83–126.
Williams, R. J. P.: The Chemistry of Lanthanide Ions in Solution and in Biological Systems. Vol. 50, pp. 79–119.

Williams, R. J. P., Hale, J. D.: The Classification of Acceptors and Donors in Inorganic Reactions. Vol. 1, pp. 249–281.

Williams, R. J. P., Hale, J. D.: Professor Sir Ronald Nyholm. Vol. 15, pp. 1 and 2.

Wilson, J. A.: A Generalized Configuration-Dependent Band Model for Lanthanide Compounds and Conditions for Interconfiguration Fluctuations. Vol. 32, pp. 57–91.

Winkler, R.: Kinetics and Mechanism of Alkali Ion Complex Formation in Solution. Vol. 10, pp. 1–24.

Wood, J. M., Brown, D. G.: The Chemistry of Vitamin B_{12}-Enzymes. Vol. 11, pp. 47–105.

Woolley, R. G.: Natural Optical Activity and the Molecular Hypothesis. Vol. 52, pp. 1–35.

Wüthrich, K.: Structural Studies of Hemes and Hemoproteins by Nuclear Magnetic Resonance Spectroscopy. Vol. 8, pp. 53–121.

Xavier, A. V., Moura, J. J. G., Moura, I.: Novel Structures in Iron-Sulfur Proteins. Vol. 43, pp. 187–213.

Zumft, W. G.: The Molecular Basis of Biological Dinitrogen Fixation. Vol. 29, pp. 1–65.

GPSR Compliance
The European Union's (EU) General Product Safety Regulation (GPSR) is a set of rules that requires consumer products to be safe and our obligations to ensure this.

If you have any concerns about our products, you can contact us on

ProductSafety@springernature.com

In case Publisher is established outside the EU, the EU authorized representative is:

Springer Nature Customer Service Center GmbH
Europaplatz 3
69115 Heidelberg, Germany

www.ingramcontent.com/pod-product-compliance
Ingram Content Group UK Ltd.
Pitfield, Milton Keynes, MK11 3LW, UK
UKHW050235080726
13610UKWH00021B/76

* 9 7 8 3 6 6 2 1 3 6 1 4 0 *